Thermal Analysis—Techniques and Applications

Thermal Analysis—Techniques and Applications

Edited by

E. L. Charsley and S. B. Warrington
Thermal Analysis Consultancy Service, Leeds Metropolitan University

ROYAL
SOCIETY OF
CHEMISTRY

Based on a course organised by the Thermal Analysis Consultancy Service in Leeds, 12–13th September 1991

Special Publication No. 117

ISBN 0-85186-375-2

A catalogue record for this book is available from the British Library

© The Royal Society of Chemistry 1992

Published by The Royal Society of Chemistry,
Thomas Graham House, Science Park, Cambridge
CB4 4WF

Printed by Hartnolls Ltd, Bodmin

Preface

The use of thermal analysis techniques has increased rapidly in the past ten years and their field of application is widening continuously. This book, which is based on a short course held by the Thermal Analysis Consultancy Service, aims to provide an overview of the principal thermal methods and their application in major areas of importance.

The instrumental topics covered include differential thermal analysis (DTA) and differential scanning calorimetry (DSC), thermogravimetry (TG), simultaneous TG-DTA/DSC, thermomicroscopy, high temperature X-ray diffraction, evolved gas analysis, thermomechanical analysis and dynamic mechanical analysis. In addition the potential of the technique of controlled rate thermal analysis is assessed.

The applications discussed in the section of the book dealing with the techniques are reinforced by more detailed discussions of some of the major fields of study in which thermal analysis has been applied. It is hoped that this inter-disciplinary approach will be of value to a wide range of scientists and engineers.

Quality assurance methods are becoming of increasing importance in research and manufacture and we are pleased to be able to include a review of the role of quality assurance in the thermal analysis laboratory.

The governing body of thermal analysis, the International Confederation for Thermal Analysis (ICTA), has played a considerable role in the development of the subject. We are therefore delighted to have an authoritative review of the activities of ICTA by the current President.

The book is concluded by a short review of sources of information in thermal analysis which will be of help to newcomers to the field.

We would like to thank all the contributors for the time and care that they have taken in preparing camera-ready scripts for this book and for their many helpful discussions. We would also like to thank our colleagues Mr A.J.Brammer and Mr J.J.Rooney for their skilled assistance. Finally we are pleased to acknowledge the editorial staff of the Royal Society of Chemistry for their help and encouragement.

E.L.Charsley & S.B.Warrington
Leeds, 1992.

Contents

List of Contributors

Professor E.L.Charsley, Thermal Analysis Consultancy Service, Leeds Metropolitan University, U.K.

Professor D.Dollimore, Department of Chemistry, University of Toledo, Toledo, U.S.A.

Mr M.J.Hardy, SmithKline Beecham Pharmaceuticals, Harlow, U.K.

Dr J.N.Hay, School of Chemistry, University of Birmingham, Birmingham, U.K.

Miss V.J.Griffin, School of Chemistry, University of Leeds, Leeds, U.K.

Dr P.G.Laye, School of Chemistry, University of Leeds, Leeds, U.K.

Dr M.Reading. ICI Paints, Slough, U.K.

Professor F.R.Sale, Manchester Materials Science Centre, University of Manchester and UMIST, Manchester, U.K.

Mr A.P.Taylor, Manchester Materials Science Centre, University of Manchester and UMIST, Manchester, U.K

Professor S.St.J.Warne, Department of Geology, University of Newcastle, Newcastle, Australia

Dr S.B.Warrington, Thermal Analysis Consultancy Service, Leeds Metropolitan University, U.K.

Mr P.H.Willcocks, ICI Materials, Wilton, U.K.

Introduction to Thermal Analysis

S. St. J. Warne

DEPARTMENT OF GEOLOGY, THE UNIVERSITY OF NEWCASTLE, SHORTLAND
NSW 2308, AUSTRALIA

1 INTRODUCTION

By definition Thermal Analysis (TA) is the term applied
to a group of methods and techniques in which a physical
property of a substance is measured as a function of temperature, while the substance is subjected to a controlled temperature programme.

Complementary determinations may measure the variations of a particular physical property at predetermined
constant temperature or pressure, conditions which are
termed isothermal or isobaric respectively.

Within the overall concept of thermal analysis, it
may be convenient to consider the terms "methods" and
"techniques" separately. Methods may be used for the
determination of a single parameter by a particular TA
method i.e. weight variations by thermogravimetry (TG).
Techniques may apply to subsequently developed specific
modifications of pre-existing methods, i.e. variable
atmosphere analysis or simultaneous thermal analysis.
Certainly a distinction should be made between these two
different ways of applying thermal analysis.

Over the last 25 years TA has passed through phases
of full recognition and consolidation to reach the
present period of further rapid increase in use and
breadth of applications ranging from polymer[1] to earth
sciences.[2]

2   FACTORS ASSISTING THERMAL ANALYSIS EXPANSION

This has been brought about by a number of factors[3] e.g.

(1). The establishment by ICTA of sets of fully certified standard reference materials.

(2). The development and increasing use of new methods e.g. emanation thermal analysis (ETA)[4] and thermomagnetometry (TM).[5]

(3). The continued upgrading of equipment, scope and refinement in application of the previously well established TA methods such as TG[6] and DTA/DSC.[7]

(4). The evolution of several new techniques e.g. variable atmosphere TA whereby the furnace atmosphere conditions may be pre-selected, maintained or changed as desired between or during TA runs.[3]

(5). The application of more than one TA method to the same sample under identical conditions to give simultaneous TA determinations. This obviates the reproducibility problems of sample and analysis conditions i.e. TG/DTA.[2]

(6). A series of complete books, specific chapters and individual or periodic reviews[8,9] on specialist aspects ranging from mineralogy,[10] polymer science,[1] fossil fuels,[7,11,12] materials characterization[6] and the excellent recent introductory overview by Brown.[13]

(7). The rapid expansion of TA fields with very marked industrial applications, for which TA methods are particularly and even uniquely suited e.g. polymer, materials and fossil fuels sciences.[1,6,7]

3   THE MAIN THERMAL ANALYSIS METHODS

The main methods listed in Table 1, are those which are the most commonly used. Clearly other low use methods in the right context may be invaluable and be considered indispensable, but these are not included.

Thermal analysis determinations are, per se, conducted under furnace atmosphere conditions of static air,

Table 1 The main thermal analysis methods

NAME OF METHOD	SYMBOL	PROPERTY MEASURED
Thermal Analysis (Group Name)	TA	Range of properties
Thermogravimetry	TG	Mass
Derivative Thermogravimetry	DTG	First derivative of mass change
Differential Thermal Analysis	DTA	Differential temperature
Differential Scanning Calorimetry	DSC	Enthalpy
Thermomechanical Analysis	TMA	Mechanical properties
Dynamic Mechanical Analysis	DMA	Frequency
Dynamic Load Thermomechanical Analysis	DLTMA	Mechanical properties
Thermodilatometry	–	Dimensions
Evolved Gas Detection	EGD	Evolved gas detection -thermal conductivity
Evolved Gas Analysis	EGA	Identity and amount of gas/gases evolved
Thermomagnetometry	TM	Magnetic properties
Emanation Thermal Analysis	ETA	Radioactive gas
Thermosonimetry	TS	Acoustic properties
Thermoparticulate Analysis	TPA	Evolution - particles
Thermooptometry	–	Optical properties
Thermomicroscopy	–	Microscopically observable properties
Thermoelectrometry	–	Electrical properties
Dielectric Thermal Analysis	DETA	Dielectric constant
Thermoluminescence	TL	Light emission
Oxyluminescence (in oxygen)	–	" "
Dynamic Reflectance Spectroscopy	DRS	Reflectance
Variable Temperature XRD	–	XRD patterns
Proton Magnetic Resonance Thermal Analysis	PMRTA	NMR determinations

at ambient pressure and with a constant heating rate. Any departure from these conditions must be clearly stated at all times.

Symbols such as P (pressure) or HP (high pressure) may prefix a method i.e. PDTA or HPTG to indicate the determinations in question have been made under above ambient furnace atmosphere pressures.

Furnace atmosphere conditions may also be dynamic, with the gas flowing through (purging) the system in order to maintain different reaction conditions round the sample under test. For example inert conditions (nitrogen) to inhibit oxidation,[14] increased partial pressure to delay and enhance carbonate reactions (carbon dioxide)[15]or to reduce iron oxides to metallic iron for detection by TM (hydrogen).[16] This technique may be referred to as "variable atmosphere thermal analysis".[2]

Further, a sequence of different gas conditions has been employed during individual TA runs, with excellent results, for a range of topics.[2] A prime example of this would be the sequence of six different gases used to compare the sulphation, regenerative sulphur dioxide and cycle efficiency of different sorbents.[17]

The related constant temperature (isothermal)[18] and stepwise (quasi-isothermal)[19] techniques have proved advantageous for specific studies for which they are particularly suited, i.e. decomposition rates,[18] mixture components and contaminants.[20] The former involves property variation measurement, against time (mass loss), at a specific constant temperature. In the latter technique the sample temperature is controlled to maintain the rate of mass loss of the sample at a pre-determined rate. The whole topic of quasi-isothermal and isobaric determinations and applications has been reviewed by Paulik and Paulik.[20]

4   STANDARDIZATION AND REPRODUCIBILITY OF INDIVIDUAL
    METHODS

A number of controllable variables affect the curves and results of TA[21,22,23] as do different TA units. It is therefore imperative to not only use standard reference materials to ensure the validity and temperature calibration of results, but also to use the same TA unit for the accurate comparison of results which must always be obtained under completely reproducible conditions.

The factors which must be standardized and reproducible are as considered under the following headings.

Sample Preparation

Solid material for most TA determinations needs to be in powdered form. This is best achieved by crushing as opposed to grinding as the latter has been shown, in a number of cases, to cause the formation of altered or even amorphous surface layers.[24,25] Being different and formed at the expense of the original material, this will give spuriously low content determinations together with additional and unexpected results, due to the new material produced.[26]

In other cases, sample size reduction may have to be undertaken under liquids, such as water or alcohol,[27] to facilitate size reduction, prevent caking or more importantly to inhibit oxidation, which is promoted by grinding in air. As a result of oxidation, an unwanted new compound (oxide) is formed in the sample, prior to TA, at the expense of some of the original sample material. Again this causes content evaluation problems.

Grain size, grain size distribution and closeness of the size fraction (degree of sorting) all have effects on the size, shape, position, resolution and even the number of peaks which result. In general terms sample materials should be crushed to pass 75 or 50 micrometre sieves, while closely sized individual size fractions are also very important for the best results.[28]

Some materials on heating will decrepitate (explode)

and as a result eject particles from the sample holder. This gives spurious TA determinations due to loss of sample material during runs, e.g. extra weight losses during TG and TM determinations. This problem, together with loss by upward attraction of grains during TM, has been countered by covering the sample in the sample holder with a quartz wool plug or a fine platinum mesh cover.[29]

Different Heating Rates and Programmes

The effect of increasing heating rates is to "telescope" reactions together causing them to occur at somewhat higher temperatures. This is because individual reactions have not had time to reach completion, or equilibrium, before the rapidly rising temperature has reached the initiation temperature of any adjacent higher temperature reactions which then commence.[28,30]

Thus, recorded reaction temperatures increase with increasing heating rates, but the reaction temperature range decreases. For duplicate DTA runs, peaks result, which although having the same area, occur with higher peak temperatures and peak heights, but narrower peak widths, thus increasing content detection limits. Conversely low heating rates (2 to 5°C) give maximum peak separation thus discouraging reaction overlap, but at the expense of poorly defined peaks.[28]

Heating rates should therefore be selected to assist in accenting the type of information required and not applied indiscriminately at a generally acceptable rate in the region of 10°C per minute.

Pre-set heating rate programmes may be used where different rates for different parts of individual runs are used, which may or may not include segments of isothermal heating.[31] These have proved particularly applicable to the elucidation of compounds in a sample, which have different decomposition rates or where determinations need to be speeded up as only certain parts of each run require the highest quality data, while other parts may show no information at all.

In contrast, controlled cyclical heating/cooling programmes may be used to monitor the efficiency, reversibility and long term dependability of chemical reactions. A good example of this is the cyclic extraction and subsequent release of sulphur dioxide from flue gases by limestone and other sorbent or scrubber materials.[17,32]

Recording or Chart Speed

The effects of chart speed are similar but the reverse of heating rates. The faster the chart speed the greater is the separation of reactions.[33]

Furnace Atmosphere Control and Effects

The TA of single compound materials or mixtures of materials, whether natural or artificial, is always complicated by the presence of more than one reaction.

(1). If separated by significant temperature differences, so that one reaction is completed before the next one starts, they should be clearly identifiable as separate entities. This case represents the easiest individual run data with which to work.

(2). However, often reaction overlap occurs, sometimes completely, so that difficulties arise as to the actual magnitude and identity of the composite peak components which may be present.

Because it is clear that certain types of reactions may be encouraged, delayed or suppressed by surrounding the sample with suitable gas conditions the use of different furnace atmosphere conditions is often invaluable.

In this way oxidation reactions may be prohibited by running in an inert gas or promoted by determinations in air or oxygen.[14] This may be achieved by passing the selected gas through the TA system and over the sample under test or actually passing it through the sample. The former method, which may be described as the flowing purge gas technique, is preferred because the latter,

although theoretically better able to preserve the
required atmosphere conditions, tends to blow some of the
sample out of the sample holder which is not acceptable.

In addition, runs made with increased partial
pressures of the same gas being given off by specific
types of decomposition reaction, delay such reactions
until the increased partial pressure of the sample
surrounding gas can be overcome. As a result the
decomposition temperature of such reactions is increased
and moves up the temperature scale to occur with
increased reaction speed. This results, for DTA, in
greater peak heights and for TG in steeper weight loss
curves, i.e. calcite in flowing carbon dioxide,[34,35]
which however leaves the decomposition curve of pyrite
unaffected.[36]

5   VARIABLE ATMOSPHERE THERMAL ANALYSIS

From the above section on "furnace atmosphere control and
effects" it has been demonstrated that TA determinations
in different and reproducible furnace atmosphere conditions represent a valuable technique. This may be conveniently referred to as "variable atmosphere thermal
analysis".[3] Using this, the furnace atmosphere
conditions may be pre-selected, maintained or changed as
desired between[37] or during[31] TA runs or preset
combinations of different purge gases[39] with dynamic
and isothermal or quasi-isothermal heating conditions may
be used.[3]

It is vital to note that the usual purge gases used
for DTA and DSC determinations from identical samples
result in peaks which show predictable variations in peak
area.[40] This is due to differences in the thermal
conductivity of these gases because, while air, nitrogen
and carbon dioxide are similar, He is approximately 6
times greater and Ar only 2/3 that of nitrogen.

6   SIMULTANEOUS AND COUPLED METHODS

With the constant aim to maintain and improve the quantification of TA methods and to obtain vital complementary
sets of information, the concept of the simultaneous
application of methods has much to offer and is becoming
widely used.

This technique involves the simultaneous application
of more than one already established TA method, to the
same sample under exactly the same conditions. Thus the
problems of reproducibility of duplicate runs, replic-
ation of samples and the constancy of experimental
variables are removed.

A distinction using the term, coupled, is made when
techniques are simultaneously applied to the same sample,
but the instruments involved are connected through an
interface.

Excellent examples of simultaneous TA are DTA-TG
where the magnitude of the DTA endothermic and exothermic
reaction peaks may be further quantified from the simultaneously determined TG weight variation curves[41] or
coupled DTA-EGA where the identity and amount of the
associated evolved gas is also determined.[35]

7   COOLING CURVES

The use of cooling curves although well documented does
not seem to be nearly as widely used as expected.

This valuable technique is only applicable to a
limited number of reactions which having taken place on
heating are reversible on cooling.

In this way, on the heating curve of mixtures many
reactions may be taking place at similar temperatures and
so interference takes place. Thus a diagnostic peak may
often go undetected due to complete or partial superposition of a larger reaction or reactions e.g. the complete
swamping of the small endothermic crystallographic
inversion peak of quartz by the large endothermic peak of
kaolinite.

However, on cooling the limited number of crystallo-

graphic reversion (quartz) or recombination reactions
(recarbonation of calcium oxide), invariably occur as
separate entities and can therefore be recognised and
measured to establish the presence of specific components
and their contents.

8   NEW METHODS

New methods continue to be produced as a result of new
technology and the requirements of research and industry.
Development initially may be for specific niches, but
expansion into other areas inevitably soon follows once
the potential of the new method becomes fully recognised.

Two examples have been selected as illustrations
viz. thermomagnetometry (TM) and proton magnetic resonance
thermal analysis (PMRTA), the very new method which was
formally launched in Sydney, Australia in June this year
(1991).

Thermomagnetometry

Thermomagnetometry may be defined as the method by
which the magnetic susceptibility of a substance is
measured as a function of temperature whilst the
substance is subjected to a controlled temperature
programme.

In the simplest terms TM equipment consists of a
thermobalance to which a magnetic field may be applied or
removed as desired during the temperature variation
programme of a sample under test. The presence of this
attractive force, on magnetic material in the sample in
the thermobalance pan (if situated below it), causes a
magnetically induced apparent mass gain on the resultant
TG curve. Such "weight gains", compared to a duplicate
run with the magnetic field absent, are directly
relatable to the amount of magnetic material produced
permanently, temporally or reversibly due to the
reaction, decomposition or magnetic transitions (Curie
points) which take place in the sample during TM runs.[42]

Determinations are however restricted to substances

containing the main Ferro-magnetic elements Fe, Co and Ni together with the rare-earth elements gadolinium, terbium, dysprosium, holmium, erbium and thulium.[43] Despite this limitation TM has proved particularly valuable in detecting and evaluating transient magnetic phases which may develop during TA. In addition applications have expanded so much recently that it has been the subject of a full scale review of the applications in the earth sciences alone.[42]

Proton Magnetic Resonance Thermal Analysis

This new method, now commercially available, is based on a benchtop NMR unit linked to a computer as described in the literature[44] and was primarily developed for coal investigations. However, it may be seen from the following description that it has wide ranging potential for application to other hydrogen bearing organic materials.

It uses signals from the protons (hydrogen nuclei) of the coal sample to determine the degree and extent of the molecular mobility. It can distinguish between fused or plastic material and the unfused or rigid material. By measuring the mobility of coal molecules as the sample is heated from 300 to 600°C, the instrument provides data to indicate the different temperature and temperature ranges at which the coal will soften, become most mobile or "plastic" and become solid again as a semi-coke (Figure 1).

A wide range of applications of PMRTA to the determination of coal properties and indicating the complementary nature of this new method to DTA, DSC and TG are extensively covered in two recent papers.[45,46]

9 RECORDING AND PRESENTING TA INFORMATION

No modern introduction to TA could be complete without reference to the very rapid applications of computerization to the data continuously being measured and recorded during thermal analyses.

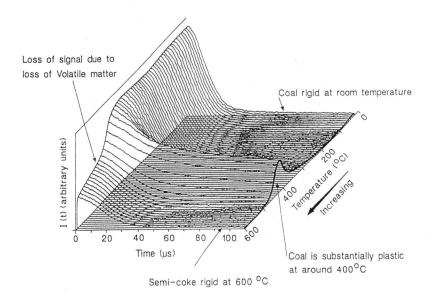

<u>**Figure 1**</u>   Stacked plot of ^{1}H NMR decay curve signals produced from a thermoplastic coal, plotted against time and temperature. From this can be obtained, as indicated, information on temperature of plasticity, loss of volatile matter and return to rigid state viz. production of semi-coke. (Reproduced with permission of CSIRO Division of Coal and Energy Technology).

Classically the results were directly and continuously registered in paper chart form by X-Y type recorders which, for example, for DTA, plotted the differential temperature against temperature to give a DTA curve. For simultaneous determinations multi-pen recorders were used to record the variations of several parameters against the common Y temperature scale. In all cases the TA data measured, if not synchronously plotted, was lost and could only be regained from a duplicate sample re-run.

With preselectable recording sensitivity and chart speed the only variable adjustments which could be made,

the scope for any further modification for the presentation of the data was essentially restricted to manual
replotting.

By comparison the application of linked microcom-
puters has extended the scope of TA determinations in the
following ways, (partly after Brown[13] and Earnest[39]):-

(1)   By the capture and processing of data as it is
        being produced.

(2)   The temporary - permanent storage of TA run data.

(3)   The display of individual run data for prelim-
        inary appraisal either as produced or later on
        demand.

(4)   Modification of the data e.g. for baseline cor-
        rection, smoothing and scaling.

(5)   The processing of the data e.g. for numerical
        definition, peak area integration and comparison.

(6)   For other calculations e.g. kinetic analysis and
        purity determinations.

(7)   The superposition of TA data and curves from
        several runs onto single diagrams for ease of
        comparison and interpretation.

(8)   For the direct production from the computer
        printer of hard copy illustrative figures and
        diagrams suitable for publication without any
        further preparation steps such as manual drafting
        or photography.

(9)   The use as per point 7, (above), of multi-run
        data obtained in different laboratories and
        transported on computer disk for inclusion and
        processing, as per point 8 (above), with data
        obtained elsewhere.

    10 THERMAL ANALYSIS COMPILATIONS

Finally, the collated, synthesized and condensed sources
of the TA information to date are invaluable and are to
be found in reference books and individual specialist
chapters in other books not devoted entirely to TA and
reviews.

A selection of such wide ranging reference publications which will collectively and specifically complement, enhance and detail modern TA methods and techniques used in a wide range of specialities and applicable from areas of basic research to economic-industrial utilization follow i.e.- general,[13,47,48] polymers [1], mineralogy,[10,49] fossil fuels,[3,7,11,12,50] DTA,[51,52] TG,[6] TM,[5] ETA,[4] and TL.[53]

Of particular relevance are the series of regular reviews which appear in Analytical Chemistry, which were initiated by Murphy, and carried on by Wendlandt and Dollimore.[8,9]

11 REFERENCES

1. E.A. Turi (Ed.), 'Thermal Characterization of Polymeric Materials', Acad. Press, New York, 1981.
2. S.St.J. Warne, Thermochim. Acta, 1991 (in Press).
3. S.St.J. Warne, Thermochim. Acta, 1990, 166, 343.
4. V. Balek and J. Tolgyessy, 'Emanation Thermal Analysis', Elsevier, Amsterdam, 1984.
5. S.St.J. Warne and P.K. Gallagher, Thermochim. Acta, 1987, 110, 269.
6. C.M. Earnest (Ed.). 'Compositional Analysis by Thermogravimetry'. ASTM Special Publ. 997, Philadelphia, 1988.
7. S.St.J. Warne and J.V. Dubrawski, J. Therm. Anal., 1989, 63, 219.
8. W.W. Wendlandt, Anal. Chem., 1984, 56, 250R.
9. D. Dollimore, Anal. Chem., 1990, 62, 44R.
10. W. Smykatz-Kloss, 'Differential Thermal Analysis: Application and Results in Mineralogy'. Springer-Verlag, Berlin, 1974.
11. K. Rajeshwar, Thermochim. Acta, 1983, 63, 97.
12. S.St.J. Warne, 'Analytical Methods for Coal and Coal Products', Clarence Karr Jr. (Ed.), Acad. Press, London, 1979, Vol.3, Chapter 52, p.447.
13. M.E. Brown, 'Introduction to Thermal Analysis', Chapman and Hall, London, 1988.
14. S.St.J. Warne, Thermochim. Acta, 1985, 86, 337.
15. S.St.J. Warne, Thermochim. Acta, 1987, 110, 501.
16. D.M. Aylmer and M.W. Rowe, Thermochim. Acta, 1984, 78, 81.
17 A.E. Duisterwinkel, E.B. Doesburg, G. Hakvoort, Thermochim. Acta, 1989, 141, 51.
18. S.St.J. Warne and D.H. French, Thermochim. Acta, 1984, 75, 139.
19. F. Paulik, J. Paulik and M. Arnold, Jour. Thermal. Anal., 1982, 25, 313.
20. F. Paulik and J. Paulik, Thermochim. Acta, 1986, 100, 23.

21. P. Bayliss and S.St.J. Warne, <u>Amer. Mineral.</u>, 1962, <u>47</u>, 775.
22. M.I. Pope and D.I. Sutton, <u>Thermochim. Acta</u>, 1978, <u>23</u>, 188.
23. J.G. Dunn, G.C. De and B.H. O'Connor, <u>Thermochim. Acta</u>, 1989, <u>145</u>, 115.
24. P.B. Dempster and P.D. Ritchie, <u>Jour. Appl. Chem.</u>, 1953, <u>3</u>, 182.
25. D.W. Clelland, W.M. Cumming and P.D. Ritchie, <u>Jour. Appl. Chem.</u>, 1952, <u>2</u>, 31.
26. S.St.J. Warne, <u>Jour. Inst. Fuel</u>, 1970, <u>43</u>, 240.
27. S.St.J. Warne and D.H. French, <u>Jour. Sedim. Petrol.</u>, 1986, <u>56</u>, 543.
28. S.St.J. Warne and P. Bayliss, <u>Amer. Mineral.</u>, 1962, <u>47</u>, 1011.
29. P.K. Gallagher and S.St.J. Warne, <u>Thermochim. Acta</u>, 1981, <u>43</u>, 253.
30. R.C. Mackenzie and B.D. Mitchell, 'Differential Thermal Analysis', R.C.Mackenzie (Ed.), Acad. Press, London, 1970, Vol.1, Chapter 4, p.101.
31. M. Ottaway, <u>Fuel</u>, 1982, <u>61</u>, 713.
32 D. Dorsey and B. Buecker, 'Compositional Analysis by Thermogravimetry', C.M.Earnest (Ed.), ASTM Spec. Publ. 997, Philadelphia, 1988, p.254.
33. G.M. Lukaszewski and J.P. Redfern, <u>Anal.Chem.</u>, 1961, <u>10</u>, Pts. I to IV., 469, 552, 630, 721.
34. S.St.J. Warne, <u>Nature</u>, 1977, <u>269</u>, 678.
35. A.E. Milodowski, D.J. Morgan and S.St.J. Warne, <u>Thermochim. Acta</u>, 1989, <u>152</u>, 279.
36. S.St.J. Warne, <u>Jour. Inst. Energy</u>, 1979, <u>52</u>, 21.
37. S.St.J. Warne and D.H. French, <u>Thermochim. Acta</u>, 1984, <u>79</u>, 131.
38. P. Bayliss and S.St.J. Warne, <u>Amer. Mineral.</u>, 1972, <u>57</u>, 960.
39. C.M. Earnest, <u>Proc. 3rd. Int. Conf. on Coal Testing</u>, Lexington, 1983, p.68.
40. S.St.J. Warne, <u>J. Thermal Anal.</u>, 1978, <u>14</u>, 325.
41. E.L. Charsley, J. Joannou, A.C.F. Kamp, M.R. Ottaway and J.P. Redfern, <u>Proc. 6th. Int. Conf. Therm. Anal.</u>, Bayreuth, Germany, W. Hemminger (Ed.), Birkhauser Verlag, Stuttgart, 1980, Vol.1, p.237.
42. S.St.J. Warne, H.J. Hurst and W.I. Stuart, (Invited Review), <u>Thermal Analysis Abs.</u>, 1988, Vol.17, p.1.
43. B.O. Haglund, <u>J. Thermal Anal.</u>, 1982, <u>25</u>, 21.
44. D.S. Webster, R. Sakurovs, L.J. Lynch & T.P. Maher, <u>Int. Conf. Coal Sci.</u>, Tokyo, 1989, p.249.
45. L.J. Lynch, D.S. Webster and W.A. Barton, <u>Adv. Magnet. Reson.</u>, 1988, <u>12</u>, 385.
46. L.J. Lynch, D.S. Webster, R. Sakurovs, W.A. Barton and P.K. Maher, <u>Fuel</u>, 1988, <u>67</u>, 579.
47. W.W. Wendlandt, 'Thermal Analysis' (3rd. Edit), John Wiley and Son, New York, 1985.
48. B. Wunderlich, 'Thermal Analysis'. Acad.Press, New York, 1990.
49. D.N. Todor, 'Thermal Analysis of Minerals'. Abacus Press, Tunbridge Wells, 1976.
50. D.D. Rustschev, (Invited Review), <u>Thermal Analysis</u>

Abs., 1989, Vol.18, p.1.
51. R.C. Mackenzie, 'Differential Thermal Analysis'.
    Acad. Press, New York, 1970, Vol.1, and 1972, Vol.2.
52. D. Schultz, 'Differentialthermoanalyse'. VEB,
    Deutscher, Verlag Der Wissenschaften, Berlin, 1971.
53. S.W.S. McKeever, 'Thermoluminescence of Solids'.
    Camb. Uni. Press, 1985.

Differential Thermal Analysis and Differential Scanning Calorimetry

V. J. Griffin and P. G. Laye

SCHOOL OF CHEMISTRY, THE UNIVERSITY, LEEDS LS2 9JT, UK

1 INTRODUCTION

The techniques of differential thermal analysis (DTA) and differential scanning calorimetry (DSC) are linked together: both are concerned with the measurement of energy changes in a substance. The word differential emphasises that measurements involve both the substance itself and a reference material. Formal definitions of the two techniques have been ratified by IUPAC.[1] From the practical standpoint the distinction lies in the instrument signal: in DTA it is proportional to the temperature difference and in DSC to the differential thermal power. Of all the thermal analysis techniques DTA and DSC are the most versatile with a range of applications which is legion. Historically mineralogy figured in the development of the techniques, more recently polymer science and pharmaceutical studies have been major areas of application. Petroleum products, hazard evaluation, assessment of energy storage systems and kinetic studies of a wide range of materials are amongst an almost endless list of applications.

DSC is usually regarded as the more quantitative technique. The area enclosed by the curve of instrument signal recorded against time is directly proportional to the energy change. DTA is one stage removed from this simple relationship. DTA is the older of the two techniques with commercial equipment appearing in the late 1940's.[2] The widespread use of both techniques can be traced to the advent of `Boersma' DTA in 1955[3] which led to heat flux DSC and to the introduction of power compensation DSC in 1964.[4] Whereas in heat flux DSC the instrument signal is derived from the temperature difference between the sample and reference, in power compensation DSC the signal is derived from the differential heat supplied to the sample and reference to maintain the temperatures the same (Fig.1).

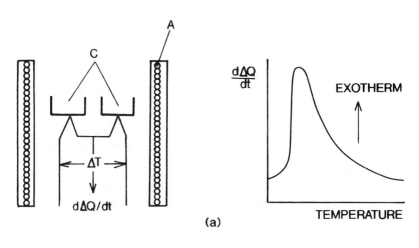

(a)

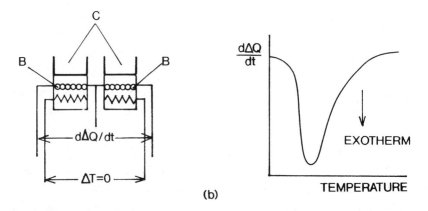

(b)

Fig.1 The principles of (a) heat-flux DSC and (b) power compensation DSC. A: furnace common to sample and reference, B: separate electrical heaters for sample and reference. C: sample and reference crucibles.

The fierce argument which ensued concerning the advantages and disadvantages of the two approaches to DSC has largely given way to recognition that it is the practical performance in a particular application rather than the theory which is of greater relevance.

The distinction between DTA and heat flux DSC is finer than might be imagined from the different nature of the instrument signal. Thus it is possible to calibrate a differential thermal analyser[5] and use it effectively as a heat flux instrument. A key factor in the development of commercial instruments is the facility to reproduce the relationship between temperature and thermal power, a prerequisite to linearisation of the instrument signal. The advent of high temperature differential scanning calorimeters is likely to reduce the manufacture of differential thermal analysers except those intended to meet specialised needs ($>1500^{\circ}$C). These differential scanning calorimeters are of the heat flux type making use of platinum/platinum-rhodium thermocouples: power compensation instruments are restricted to a maximum temperature of about 750°C.

2 INSTRUMENTATION

Since the introduction of commercial differential scanning calorimeters great strides have been made in their instrumentation. There has been a drive for increased sensitivity and the increase by an order of magnitude has led to sample masses often of less than a mg. Paradoxically there is a need for instruments which will allow the study of several grams of sample where the heterogeneous nature of the material makes the use of small quantities disadvantageous. Different temperature requirements ranging from -150°C to 1500°C are met by a variety of equipment. The uncertainty in calorimetric measurements using differential scanning calorimeters is usually claimed to be about 1-2%. Modern equipment is designed for ease of use. A considerable emphasis has been placed on fast heating ($>100^{\circ}C$ min^{-1}) and cooling. The latter ensures a fast through-put of samples, a requirement which may be of paramount importance in some industrial situations. Equipment has been designed to allow the study of more than one sample at a time but it is with robotic systems that measurements may be made more automatic. Close control of the environment of the sample is a feature of modern instruments and with some equipment there is the option of working at high or low pressures.

The last decade has seen the increasing use of computer systems both to control experiments and interpret the results. Some new differential scanning calorimeters have been introduced over the last year or so but the main preoccupation has been the design of software. Undoubtedly major improvements have been made but inevitably computations are being performed on data that are becoming increasingly distanced from the raw instrument signal. Also inevitable is the incompatibility which seems to exist between different versions of software: it has been rumoured that differing results can be obtained with different versions of software. If this is so it raises a doubt as to exactly what is being calculated. A comment on the UK instrument market which is noteworthy is the appearance of Japanese equipment in a market hitherto dominated by instruments made in the USA, UK and Europe.

3   ANALYSIS OF THE THERMAL ANALYSIS CURVE
Quality control has always been an important use of DTA and DSC. The claims made for ease in the use of equipment and in the interpretation of results are highly relevant where the emphasis is on comparisons between results. A major concern of DTA has always been the shape of thermal analysis curves which may often be used for qualitative analysis. However it is the quantitative application which attracts the most attention in the scientific literature. Key features of

the thermal analysis curve are signal displacement and peak area which lead to heat capacity (Fig.2) and enthalpy respectively. Analysis of the shape of fusion curves leads to information on purity, an important application of thermal analysis (Fig.3). More contentious is its use to obtain kinetic information. Claims of simplicity are a disservice since they mask the complexities of the techniques.

Wendlandt[6] has listed some sixteen characteristics which influence the shape of thermal analysis curves: some are incorporated in the design of the apparatus, others are available to the operator. Sample mass and preparation, crucible material, heating rate, furnace atmosphere, all of these are known to affect the shape of the results, often profoundly.

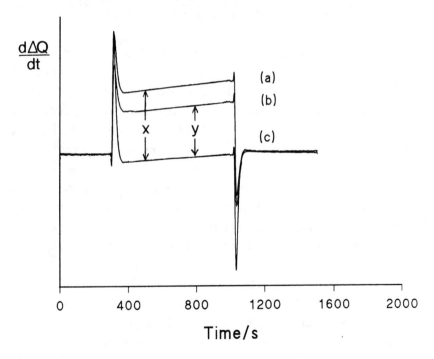

Fig.2 The measurement of the heat capacity (C) of an oil. The thermal analysis curves are for (a) sapphire standard (b) oil and (c) empty pan. $C_{oil} = (C_{sapphire} \cdot m_{sapphire} \cdot y) / (m_{oil} \cdot x)$, where m denotes the mass of the sample.

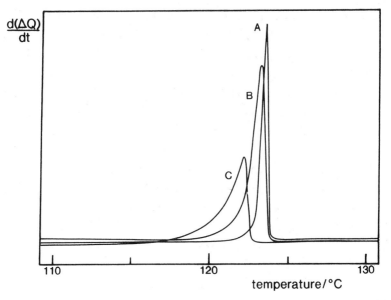

Fig.3 Power-compensation DSC curves for benzoic acid of varying purity: (a) 99.9%; (b) 99%; (c) 97%.

The effect of sample containment on the results for an emulsion explosive is shown in Fig.4. The thermal analysis curve for a sample studied in an open crucible is dominated by a large endotherm! In a closed crucible the expected exotherm is obtained. The effect of heating rate on the shape of DSC peaks is shown in Fig.5. The movement of the peaks to higher temperatures as the heating rate is increased is clearly discerned. Also highlighted is the very different appearance of peaks recorded against temperature and time. Only in the latter case are the areas independent of the heating rate.

A theoretical model for DTA and DSC depends on a description of the temporal and spatial distribution of heat in the sample and its environment in the sample holder. Inevitably assumptions must be made: the effect of these assumptions may be minimised in the design of the equipment.

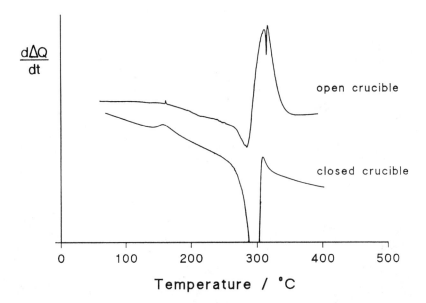

Fig.4   DSC curves for an emulsion explosive recorded with the sample in (a) an open crucible and (b) a closed crucible. (Endotherms are shown upwards.)

The theory set out by Gray[7] in 1968 which remains the most quoted has the advantage of conceptual and mathematical simplicity. It accounts for the shape of thermal analysis curves albeit by making sweeping assumptions. Since then there have been many publications dealing with theoretical models of instrument behaviour, both thermal and electrical analogues. These are exemplified by Refs.8-12. These allow aspects of instrument behaviour to be investigated although often as a theoretical exercise with limited practicality. It remains true that DTA and DSC are rooted in the empirical.

4   CALIBRATION

Calibration lies at the heart of the techniques.[13,14] Unlike classical calorimetric techniques which employ electrical heating to relate results to international primary standards, thermal analysis is dependent on chemical standards. There is a tendency to lose sight of

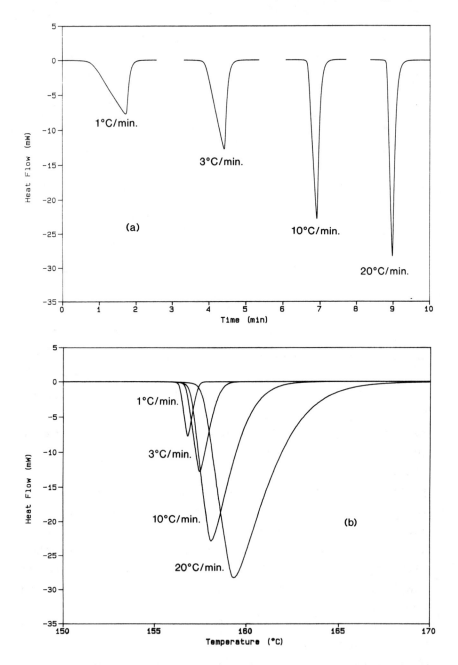

Fig.5   DSC curves for the fusion of 10mg of indium at
various heating rates recorded against (a) time and
(b) temperature.

the obvious fact that results cannot be more accurate or have greater precision than the calibration. The production of an unwarranted number of significant figures is an unfortunate characteristic of results produced by computers. The importance of calibration has been recognised by the International Confederation of Thermal Analysis (ICTA) which has established a new committee to consider standards. With the advent of modern instruments it is unlikely that the temperature standards promoted by ICTA in 1974[15,16] and available from the National Institute of Science and Technology (NIST) will be used in the way originally envisaged. The object was to ensure that instruments were calibrated to common standards. The uncertainty in some of the certificated temperatures is no longer acceptable in the use of present day equipment with its much greater resolution. Nowadays temperature calibration is achieved by using pure metals and ascribing the extrapolated onset temperature of the fusion peak to the melting temperature. The relevance of thermal lag and the issue of extrapolating temperatures to zero heating rate have been discussed.[17,18] A related problem is how the calibration may be used for samples with thermal properties different from the calibrants.

Calibration for enthalpy presents a very confused situation; the more so since the National Physical Laboratory has withdrawn indium as a standard calibrated sample. The position was summarised in a recent publication[19] where it was made clear that there is a need to resolve the entire question of calibration. The National Physical Laboratory has available organic materials with well defined enthalpies of fusion[20] but the temperature range is limited to < 250°C. There is an urgent requirement for new measurements by classical calorimetry on metals. Work is ongoing to produce new values. In this context values measured by thermal analysis do not constitute an independent determination. At the highest level of precision there must be some question regarding the use of results reported for isothermal conditions in techniques where the temperature is changing. There remains the nagging doubt regarding the validity of measurements for exothermic events carried out on the basis of endothermic calibrations.

The use of sapphire in heat capacity measurements is well established[21] and provides a calibration of the signal displacement. Area measurements may be made to obtain an enthalpy calibration but the areas are small by comparison with those for fusion endotherms. Sapphire is one of the few calibrants available for high temperature DSC, a technique which has generated considerable interest in recent years. It places great demands on both the instrument manufacturer and the skill of the operator. The calibration may be relevant to subsequent measurements of heat capacity but its transferability to other determinations is not assured.

5 APPLICATIONS

The last decade has been marked by a growing interest in linking DTA and DSC with other techniques to enhance to information to be gained: the use of DTA and thermo-gravimetry in combination has long been established. The advent of photo-DSC[22] promises a new source of information in the study of curing reactions. DTA and DSC continue to be used in specialised applications. The measurement of thermal conductivity has attracted interest over a number of years. An apparatus used in connection with the development of the theory of the technique[23] is shown in Fig.6. The design of some equipment represents a compromise between DSC and classical calorimetry. The Setaram C80 heat flux calorimeter may be calibrated electrically and has been used for an independent measurement of the enthalpy of fusion of bismuth. The larger reaction vessels of this calorimeter allow more complex experiments to be mounted but the heating rates are limited by the larger thermal

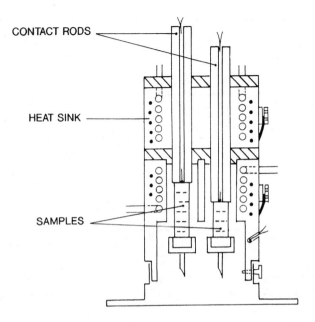

CONTACT RODS

HEAT SINK

SAMPLES

Fig.6   An apparatus for the measurement of thermal conductivity.

inertia. The drive to establish recommended procedures[24-26] based on the use of DSC has continued but the unavoidable fact remains that for some determinations different laboratories often produce different absolute values although the results show the same trends.[27] A recommended procedure for testing high alumina cements was published in 1975 by The Thermal Methods Group of the Royal Society of Chemistry. (Fig.7)

There is a regularly recurring interest in the more fundamental aspects of the techniques. Recent work[28,29] has questioned the operation of power compensation DSC although the results illustrate the remarkable resolution of modern equipment. The thermodynamic aspects of the techniques seem to attract less attention even though they provide the basis for calculating results.[30] Kinetic measurements continue to be a force of interest. Over the years there has been an unfortunate emphasis on the single experiment as a

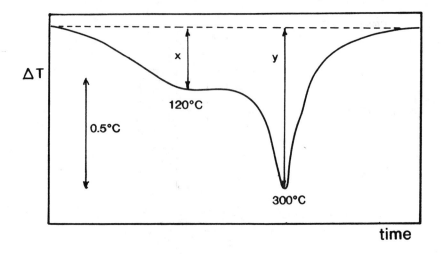

Fig.7 DTA curve for a high alumina cement sample from
a test drilling. Degree of conversion = 100y/(x+y).

source of kinetic parameters. This fails to recognise
the complexity of reaction kinetics. More appropriate
might be the use of the techniques to obtain the triplet
of information, reaction rate, extent of reaction and
temperature for subsequent analysis by well proven
classical methods. Some authors[31,32] have underlined the
need for supporting evidence before ascribing a reaction
mechanism. DSC and DTA are macroscopic techniques which
cannot give information about microscopic processes
except by inference.

6 CONCLUSIONS

There can be no doubt regarding the interest surrounding
DTA and DSC. It is rare to find techniques which
encompass such a wide range of applications. Often they
provide information difficult or impossible to obtain by
other methods. Even so there is a certain obscurity
surrounding their quantitative application. It is this
aspect perhaps which accounts for the continuing
fascination these techniques evoke.

7 REFERENCES

1. R.C. Mackenzie, Nomenclature for Thermal Analysis IV (Recommendations 1985), Pure Appl. Chem., 1985, 57. 1738.
2. R.C. Mackenzie, Thermochim. Acta, 1984, 73, 307.
3. S.L. Boersma, J. Am. Ceram. Soc., 1955, 38, 281.
4. E.S. Watson, M.J. O'Neill, J. Justin and N. Brenner, Anal. Chem., 1964, 36, 1233.
5. Feng Hongtu and P.G. Laye, Thermochim. Acta, 1989, 153, 311.
6. W.Wm. Wendlandt, `Thermal Analysis', 3rd Edition, Wiley, New York, 1986, p. 228.
7. A.P. Gray in `Analytical Calorimetry', R.F. Porter and J.M. Johnson (Eds), Plenum, New York, 1968, p. 209.
8. J.H. Flynn in `Status of Thermal Analysis', O. Menis (Ed), N.B.S. Special Publication, 1970, 338, 119.
9. P. Claudy, J.C. Commercon and J.M. Letoffe, Thermochim. Acta, 1983, 68, 305.
10. S.C. Mraw, Rev. Sci. Instrum., 1982, 53, 228.
11. Y. Saito, K. Saito and T. Atake, Thermochim. Acta, 1986, 99, 299.
12. F.W. Wilburn, D. Dollimore and J.S. Crighton, Thermochim. Acta, 1991, 181, 173, 191.
13. ASTM Standard Practice for Temperature Calibration of Differential Scanning Calorimeters and Differential Thermal Analysers, E967-83 (reapproved 1987). Annual Book of Standards 14.02:658-661, 1991.
14. ASTM Standard Practice for Heat Flow Calibration of Differential Scanning Calorimeters, E968-83 (reapproved 1987). Annual Book of Standards 14.02: 662-665, 1991.
15. H.G. McAdie in `Thermal Analysis' Vol. 1, H.G. Wiedemann (Ed), Berkhauser Verlag, Basel, Switzerland 1972, p. 591.
16. H.G. McAdie in `Thermal Analysis' Vol. 1, I. Buzas (Ed), Akademiai Kiado, Budapest, 1975, p. 251.
17. M.J. Richardson and P. Burrington, J. Thermal Analysis, 1974, 6, 345.
18. G.W.H. Höhne, H.K. Cammenga, W. Eysel, E. Gmelin and W. Hemminger, Thermochim. Acta, 1990, 160, 1.
19. J.E. Callanan, S.A. Sullivan and D.F. Vecchia, J. Res. Nat. Bur. Stand., 1986, 91, 123.
20. R.J.L. Andon and J.E. Connett, Thermochim. Acta, 1980, 42, 241.
21. M.J. O'Neill, Analyt. Chem., 1966, 38, 1331.
22. R. Sastre, M. Conde, F. Catalina and J.L. Mateo, Rev. Plas. Mod., 1989, 57, 375.
23. T. Boddington and P.G. Laye, Thermochim. Acta, 1987, 115, 345.
24. ASTM Standard Test Method for Arrhenius Kinetic Constants for Thermally Unstable Materials, E698-79 (reapproved 1984). Annual Book of Standards, 14.02: 520-526, 1991.

25. ASTM Standard Test Method for Mole Percent Impurity by Differential Scanning Calorimetry, E928-85 (reapproved 1989). Annual Book of Standards 14.02: 648-651, 1991.
26. ASTM Standard Practice for Calculation of Hazard Potential Figures-of-Merit for Thermally Unstable Materials, E1231-88. Annual Book of Standards, 14.02: 775-779, 1991.
27. A.A.J. Cash and P.G. Laye, Analyt. Proc., 1985, 22, 43.
28. R.M. Flynn, J.H. Flynn and T.J. Bent, Thermochim. Acta, 1988, 134, 401.
29. G.W.H. Höhne and E. Glöggler, Thermochim. Acta., 1989, 157, 295.
30. T. Boddington and J.F. Griffiths, Fundamental Aspects of Differential Scanning Calorimetry, Proc 7th Symp. Chem. Probl. Connected Stabil. Explos., Smygehamn, Sweden, 1985, p. 217.
31. P.D. Garn, Thermochim. Acta, 1987, 110, 141.
32. M.E. Brown, Thermochim. Acta, 1987, 110, 153.

Thermogravimetry

D. Dollimore

DEPARTMENT OF CHEMISTRY, UNIVERSITY OF TOLEDO, TOLEDO, OH USA

1 INTRODUCTION: THE DEVELOPMENT OF THERMOGRAVIMETRY

Honda[1] is usually credited with the invention of thermogravimetry and the use of the name "thermo-gravimetric balance". He was probably led to this kind of work because he was also actively engaged in using a similar balance in his studies on magnetochemistry and the conversion amounted to removing the magnet around the sample and replacing it with a furnace.[2] To make possible Honda's construction of the thermobalance various other developments had to occur. Joseph Priestley (1733–1804) in the eighteenth century was still advancing the Phlogiston theory and the notion of "negative weight" which would have proved a setback to further advance had not his ideas, (but not his experiments) been speedily corrected by Lavoisier.

The experiments on chemical reactions by Joseph Black[3] and by Bryan Higgins[4] largely on alkaline earth salts indicated the importance of obtaining information by weighing materials before and after heating.

These men were hampered by lack of a suitable temperature scale and the method of measurement. They had available liquid in glass thermometers based on the Fahrenheit scale or the Celsius with its centigrade scale. The concept of the absolute zero was introduced by Lord Kelvin.[5] The discovery of the thermoelectric effect by Seebeck[6] in 1822 formed the basis of the use of thermocouples. The use of such devices as reliable thermometers was delayed because of difficulties in obtaining the required metals and alloys to the required purity and the narrow specifications required. Nevertheless Becquerel[7] and Pouillet[8] showed the use of such devices in measuring temperature. Le Chatelier[9] showed that the thermocouple could be used as a really accurate temperature measuring device. Resistance thermometers and the optical pyrometer were developed later[10,11] but in present day TG equipment the temperature measuring and controlling device is most often the thermocouple.

The use of balances dates back to antiquity. Micro-balances are balances designed to weigh, with great sensitivity and precision, quantities of materials in the micro-gram range.[12] One of the earliest designs for such a balance was described by Warburg and Ihmori.[13] In most descriptions of micro-balances five types are noted:

    i)   the displacement type (beam and cantilever)

    ii)   the torsion displacement type

    iii) the beam-knife-edge type

    iv) the spring balance

    v)   restoration type balances.

Nernst and Riesenfeld in 1903 described a torsion-displacement type of microbalance[14] which was used to determine mass loss on heating Iceland spar, opal and zirconia. Around this time the beam-knife-edge type balance was also developed as a micro-balance.[15,16] Subsequent developments of these balances were for use in vacuum, for adsorption studies and for heat treatment under controlled atmospheres. Later developments also included designs which could handle heavier loads with easy detection of small mass changes in the microgram range.

Thus Honda in inventing the thermobalance had all the components to hand. Others had come close but failed to use the equipment in the manner associated with today's usage. Thus Brill[17] used a Nernst microbalance and used a furnace for heating, removing the sample at

short and regular intervals to weigh it on the
microbalance. Urbain constructed an apparatus consisting
of a conventional balance modified for null-point
electromagnetic compensation with the sample hanging from
the balance arm in an electric furnace. For some reason
its development was not extended to reliable use. It must
however be regarded as the first thermobalance. The
importance of Honda's achievement was simply that his
equipment was reliable and reports of its use in solving
problems appeared at the same time with many more
appearing in the period after that first important
publication.

2 DESIGN FEATURES

The design of the thermobalance has settled into clearly
recognizable features, namely a good precision balance,
a furnace capable of being programmed and a work station
computer that will control the equipment and process the
data. In the last few years the emphasis on change has
centered around the work station. The older texts will
describe a recorder from which the data will be plotted
as mass change against temperature and any subsequent
calculations assembled from the plot and calculated
separately. Now with the computer work station the
equipment can be controlled to produce the most
sophisticated temperature regimes, control the rate of
decomposition, and process the data in any number of
ways. The programs can be bought commercially, at the
time the equipment is bought, or programs added to
existing work stations by commercial software firms, or
can be down loaded into a second computer for processing
according to ones individual requirements. The field here
is still developing. The two basic requirements that
remain are the balance itself and the furnace.

The Balance

Most commercial balances used for this work are null
point balances. The balance must be equipped with a

sensor to detect the deviation of the balance beam from its null position. Various methods can be used to indicate this deviation and to restore the balance to the null position. Two methods are usually adopted to detect the deviation from the null position, namely optical methods and electronic methods.

In the optical methods, use is made of a photoelectric cell. The usual equipment is a light source, a shutter or a mirror and either single or double photoelectric cells. The movement of the shutter caused by the deviation of the balance beam from the null position intercepts the light beam. This alters the light intensity reaching the photoelectric cell. A current is produced from these cells which varies according to the light intensity received and this is used to restore the balance to its null position. In electronic methods the deviation of the balance beam is detected by various methods such as the change in capacity of a condenser as a function of balance beam displacement, the variation in inductive coupling between plates and coils, magnetic coupling, an alteration of the output of a differential transformer as a function of armature displacement, a variation in nuclear radiation flux or a difference in output of a strain gauge circuit. This list is not complete but illustrative of the wide variety of methods adopted. The deflection of the balance and its extent is found by any one of these methods and the signal generated is then used to restore the balance to the null position. Thus in a system using a solenoid coil operating on a magnetic armature attached to the beam of the balance and freely suspended within the coil a current can be applied to the coil of sufficient magnitude to bring the balance back to the null position. The magnitude of the current is decided by the signal generated in the detection device. The equipment has to be calibrated so that the mass of material or mass change can be generated as a signal on the computer work station.

The main tendency in balance design has been to use a microbalance so that small samples can be used. This means that the errors due to temperature gradients throughout the sample are minimized. However, modern commercial equipment design means that operation of the TG unit must be carried out at one atmosphere but the kind of gas can be altered and it is easy to control the rate of flow of the gas over the sample. The use of flowing gas is to be generally recommended as otherwise the gas produced in the reaction might not be cleared away from the source of the decomposition process. There are firms, however, who make special thermobalances, designed to take large loads and to resist corrosion from reactive gases produced in the decomposition. However, no commercial units are available which enable TG to be practised at vacuum or at controlled low pressures. The balances, however, are available commercially which can be operated at vacuum or at controlled low pressures up to one atmosphere. The user is then obliged to construct a thermobalance by buying these specialized balances together with a suitable furnace. There are also balances available which can operate at pressures up to very high atmospheric pressures and these with care could be adapted for TG investigations with a controlled temperature regime.

The Furnace

There are certain features which are necessary for furnaces used in TG units. These are as follows:

1. The furnace should be able to reach about 100-200°C above the desired working temperature.

2. A hot zone in the furnace should be of reasonable size to accommodate the sample and crucible so that the sample can be held at a constant uniform temperature.

3. The furnace should be wound non-inductively to avoid possible errors in weighing.

4. The temperature regime should be faithfully

reproduced. The computer work station is capable of a wide variety of temperature programs. A plot of temperature against time should be available to check that the control is accurate.

5.   Radiation and convection currents from the furnace should not affect the weighing system.

6.   The recorded temperature should be the sample temperature.

7.   If operated in the presence of a corrosive atmosphere the linings of the furnace must be made of material capable of resisting chemical attack.

Furnaces are rarely constructed by research workers *in situ* but a general knowledge of the winding material is necessary as some high temperature windings can only be used in particular atmospheres. Commercial equipment can go as high as around 1650°C whilst most decomposition studies only require a top temperature of less than 1000°C. The size of the furnace must also be considered. Furnaces of low-mass cool quickly but hold little heat whereas the converse holds for high-mass furnaces. If using large mass furnaces it is often advantageous to buy two furnaces so in heavy usage of equipment one can be cooled down with a fan whilst the other is in use. The disadvantage of a low-mass furnace may be that the temperature program may be technically more difficult to manage because of "overshoot" and similar problems. However a high-mass furnace may hold an isothermal temperature but can take a considerable time to achieve the required temperature. In kinetic studies it is usually desirable to reach the isothermal temperature in as short a time as possible, which can best be achieved by using a low-mass furnace.

The position of the furnace is important with respect to the balance assembly. Typical variations are shown in Figure 1. Sometimes as in the spring balance the position of the furnace is dictated by balance design. In recent innovations identical furnaces are placed on both balance arms to minimize the effect of convection currents.

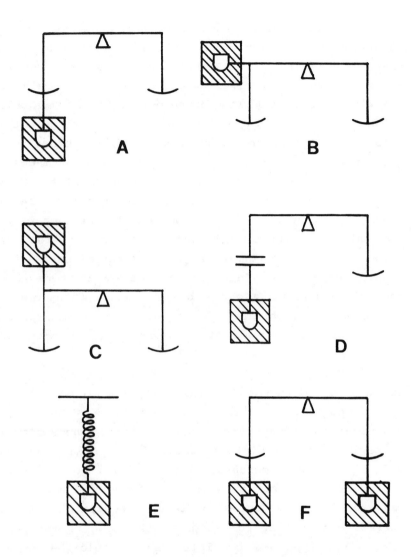

<u>Figure 1</u>   Schematic illustration of positions of the furnace with respect to the balance : A. Furnace below balance, B. Furnace at side of balance, C. Furnace above balance, D. Remote Coupling, E. Spring Balance, F. Twin Furnaces.

The most common method of measuring temperature in TG balances is a thermocouple. The choice is made by the instrument manufacturer. Resistance thermometers are used in some instances. There should be some method of checking on the temperature. One method is to use a solid "link" of a pure metal which when it melts slips either onto or off the balance crucible.[18] Another method is to use a ferromagnetic material suspended within a magnetic field which gives an anomalous mass reading at the Curie temperature.[19] The position of the thermocouple is important. As noted it should be near to the sample. The instrument designer may use this to control the temperature of the furnace as well as the measurement of sample temperature. In other circumstances of design a control temperature might be built into the furnace walls and then it might be advisable to have a separate thermocouple either embedded in or near to the sample.

3   DATA PRESENTATION

In the field of thermal analysis, organizations, national and international are in existence and these organizations have set out certain nomenclature abbreviations, definitions and standards. The suggestions made include recommendations concerning reporting data. Equipment varies and to a certain degree different data may result from using different equipment. It is therefore important to designate the type of equipment used. The type of crucible used should be stated including the type of metal, its shape and size. The temperature regime imposed on the sample should be clearly noted. The atmosphere over the sample should be noted in terms of pressure, composition and purity. It should also be recorded whether the atmosphere is static or dynamic and if the latter which flow rate was used.

Other recommendations regarding reporting data refer to the fact that samples under investigation are most often solids. The most important feature in this case is the fact that to specify the condition of a solid phase it is necessary to record its prehistory. Thus a sample of calcite may vary in its decomposition details according to its crystal size. The sample should also be identified by definitive name, empirical formula or equivalent compositional data, and purity.

The actual TG data should initially be recorded as mass plotted against the temperature regime details. In the case of linear heating, the heating rate should be noted and the plot is then of mass against temperature. Additional plots include mass loss, fraction decomposed, percent mass loss or molecular composition, all plotted against the temperature.

With the advent of the computer work station it has become increasingly possible to obtain a derivative plot (DTG).

Simple TG plots are shown schematically in Figure 2. This indicates the important features of the curve. The DTG plot may show these features more distinctly. Such curves are schematically shown in Figure 3 with a suggestion of details which should be noted. These details are especially helpful in determining the kinetic features of the decomposition process.

Calcium oxalate monohydrate seems to be an excellent choice to test performance of a TG unit. Three distinct stages of decomposition can be discovered and although the temperature may vary according to the conditions and equipment design the percent loss for each stage is a good test of the balance assembly. The TG data for this material is given in Figure 4 from data supplied by Evans.[20] Each stage corresponds to reactions as follows:

$$1st\ Stage \quad CaC_2O_4 \cdot H_2O(s) \xrightarrow{\text{endothermic}} CaC_2O_4(s) + H_2O(g)$$

$$2nd\ Stage \quad CaC_2O_4(s) \xrightarrow{\text{endothermic}} CaCO_3(s) + CO(g)$$

$$3rd\ Stage \quad CaCO_3(s) \xrightarrow{\text{endothermic}} CaO(s) + CO_2(g)$$

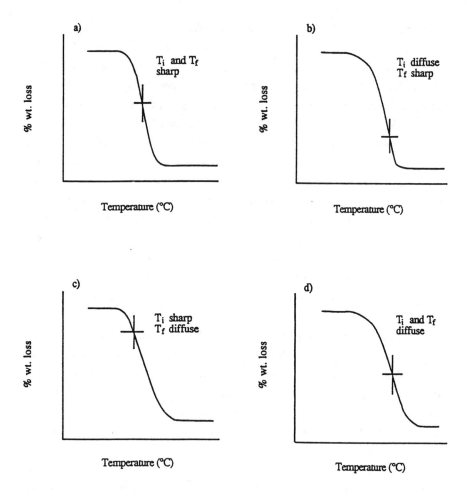

<u>Figure 2</u> Schematic representation of TG plots : typical
shapes and point of maximum slope (i.e. $(d\alpha)/(dT)$ max),
T_i onset or initial temperature. T_f final temperature.

The second stage is sometimes found to be exothermic
if a suitable catalyst is present for the oxidation of CO
to CO_2 to occur in the presence of air.[21]

$$2CO + O_2 \xrightarrow{\text{catalyst}} 2CO_2$$

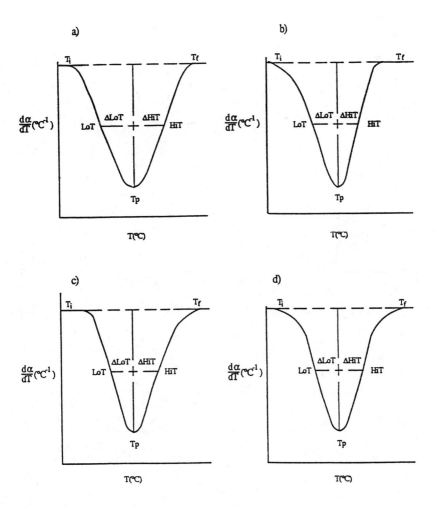

<u>Figure 3</u> Typical DTG plots corresponding to TG plots shown in Figure 2

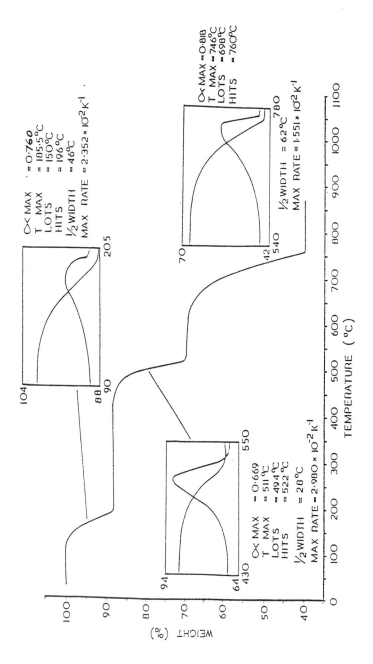

Figure 4   Shape parameters for the decomposition of calcium oxalate monohydrate at 10 deg/min., under an atmosphere of air (supplied by T.A. Evans).

4 EFFECTS OF EXPERIMENTAL ENVIRONMENT

It is often pointed out that TG is a technique where the results are going to vary with the equipment used and the design of the equipment. However this puts the cart before the horse! The fact is that in 9 applications out of 10 the experimenter is dealing with a solid. Now as already pointed out the solid phase is the only phase where one has to consider the prehistory. It is easy enough to point out that various crystallographic forms of silica can be noted, i.e. quartz, crystobalite and tridymite, or the different crystalline states of calcium carbonate, i.e. calcite, aragonite, vaterite, etc. All these forms co-exist at room temperature, transition temperatures are quoted in the literature but the transformations may be accompanied by significant kinetic factors.[22,23] The process of quenching to freeze a high temperature phase so that it co-exists at lower temperatures is a recognition of the fact that the pre-history must be noted. The pre-history must be noted because variation in particle size, size of crystals, and type of surface cause differences in solid state activity. However just to note that definite crystal forms of the same material may provide samples of various reactivity is to ignore the fact that many solids exist with a disordered long range arrangement of constituent species. Such solid phases are termed amorphous. The term is ill-advisedly considered by some to be synonymous with the vitreous state. However this term is better reserved for a super-cooled liquid, a phase in which long term order is also absent. Both the vitreous state and the amorphous are meta-stable. The important aspect of all investigations of solid state decompositions is that the reactivity is located at a surface. In the case of gasification reactions such as carbon blacks reacting with oxygen,

$$C + O_2 \rightarrow CO_2 \quad \text{and} \quad 2C + O_2 \rightarrow 2CO$$

the reaction interface can probably be identified with

the surface area identified by adsorption of gases.

In the case of limestone decomposition

$$CaCO_3 \rightarrow CaO + CO_2$$

the reaction interface, its formation and manufacturing treatment decides the reactivity of the solid. It is because the solid phase is most often the subject of TG investigations that it is so necessary to quote the experimental conditions.

The Solid State

It is worthwhile examining solid state structures further and to appreciate that the surface structure dictates the solid state reactivity observed in TG experiments. Conventionally the usual approach to the solid state is to state structures and construct equilibrium diagrams of condensed systems. This ignores the fact that the surface or the reaction interface structure may be quite different from the bulk structure and that it is the extraordinary variation in the surface area that occurs in a material that otherwise can be identified by the same chemical formula that makes it necessary to very carefully record the experimental conditions of a TG run.

X-ray studies are increasingly available for elucidation of crystal structures with the introduction of the newest available equipment. The use of such data to elucidate solid state reaction mechanisms is a naive exercise. This is because the reaction takes place either at the reaction interface (as in calcium carbonate decomposition) or in the liquid phase (as in the degradation of many nitrates or some polymer systems).

The Surface State

The structure at the interface of the solid phase is represented by a distorted structure reflecting the unbalance of forces at the surface and the polarizability of the species appearing at the surface. Thus in most oxides the real surface is represented by an array of oxide ions (witnessed by a similarity in the heats of wetting of such oxides with respect to unit surface area).

This unbalance of forces at solid surfaces leads to alteration both in the extent of the surface (an extensive property) and the surface energetics associated with the surface (intensive). This results in sintering of the solid particles during decomposition. The effects of sintering causing a reduction in surface area are counteracted by thermal activation brought about by the shattering of particles due primarily to a difference in density (and hence volume) between solid reactants and solid products.[24] Both sintering and the activation process have rates which are temperature dependent.[25] It would seem appropriate then in recording the surface parameters of at least the starting initial solid reactant material in terms of particle size, particle size distribution or surface area.

Effect of the atmosphere

The environmental atmosphere around the sample can cause drastic changes in TG results. If the environmental atmosphere is the product gas then very often an increase in pressure of this gas can cause an increase in the temperature of the decomposition. Thus the temperature of calcium carbonate decomposition is raised on the presence of increasing pressures of carbon dioxide. This is an example of the operation of Le Chatelier's principle. There are examples where the decomposition temperature does not respond to the pressure of carbon dioxide in the same manner. Thus in the first stage of the decomposition of dolomite ($CaMg(CO_3)_2$) the temperature of decomposition may actually decrease by a few degrees as the pressure of carbon dioxide is increased. This first stage is generally written

$$(CaMg(CO_3)_2) \rightarrow CaCO_3 + [MgCO_3]$$

The $MgCO_3$ is produced at a temperature above its normal decomposition temperature so that it decomposes as formed;

$$MgCO_3 \rightarrow MgO + CO_2$$

The remaining calcite dissociates at a higher temperature

$$CaCO_3 \rightarrow CaO + CO_2$$

and is dependent on the pressure of CO_2 as expected by consideration of Le Chatelier's principle.

Other effects of the environmental gas can be seen by considering the thermal decomposition of oxalates.[26] Nickel oxalate dihydrate decomposes in two ways depending on the atmosphere. In nitrogen;

1st Stage

$$NiC_2O_4.2H_2O \xrightarrow{\text{endothermic}} NiC_2O_4 + 2H_2O$$

2nd Stage

$$NiC_2O_4 \xrightarrow{\text{endothermic}} Ni + 2CO_2$$

In air or oxygen this last step is;

$$NiC_2O_4 \rightarrow [Ni] + 2CO_2 \text{ and then as formed;}$$

$$2Ni + O_2 \rightarrow 2NiO$$

and the overall reaction is exothermic, and the mass change is quite different.

This has to be contrasted with zinc oxalate dihydrate. On decomposition in nitrogen this is;

1st Stage

$$ZnC_2O_4.2H_2O \rightarrow ZnC_2O_4 + 2H_2O$$

2nd Stage

$$ZnC_2O_4 \xrightarrow{\text{endothermic}} ZnO + CO + CO_2$$

In air or oxygen the reaction is the same but is

exothermic. The reason for this is that zinc oxide is a good catalyst for the reaction

$$CO + \tfrac{1}{2}O_2 \xrightarrow{\text{ZnO catalyst}} CO_2$$

Other instances can be found concerning the influence of the experimental atmosphere - some of them important industrially. Thus coal catches fire in air but converts to coke in nitrogen. The same is true of composites of poly(styrene-butadiene) and carbon. However the different behavior can be used to give a compositional analysis subject to some reservations. This is shown schematically in Figure 5.

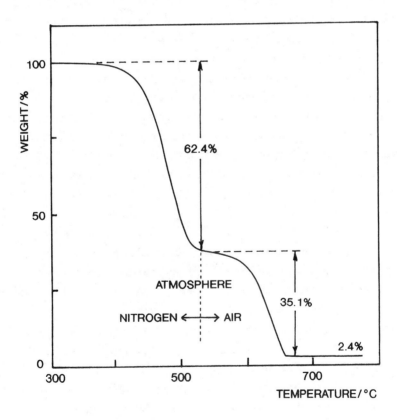

Figure 5 TG curve for an elastomer system containing Carbon Black in nitrogen up to 530°C and air beyond that (supplied by P.Manley).

The degradation of the polymer can be ascertained from a first TG run in nitrogen and the carbon by gasification when the atmosphere is changed to air as shown in the Figure.   Care   must   be   taken   however,   for   recent experiments have shown that for certain polymers which normally depolymerize on heating when in contact with certain solids that part of the polymer adjacent to the solid surface degrades by chain stripping to give a carbon residue.[27]

5 THERMODYNAMIC FACTORS

TG data relates to mass loss at various temperatures. The effect of the pre-history of the solid phase means that the material may be removed significantly from its equilibrium state. Of course in many processes the rate of reaction is significant and an end product form can be the form stable at that temperature. In such instances the "classical" thermodynamic treatment appears to fit, but often with one exception, namely that due to the location of the process at the reaction interface, the reverse of the process may become impossible. Thus in the decomposition of calcium carbonate, the recarbonation of the oxide can be achieved but in the decomposition of magnesium carbonate the recarbonation of the oxide is very difficult and may be confined to the surface layers only.

In polymers one method of dealing with the pre-history of a solid phase is to eliminate the history of the sample. The method generally adopted is to simply take the sample temperature up to produce the liquid state and cool back down to the solid phase region. The history can then accurately be built into the solid phase by controlling the cooling.

In spite of the points made above a surprising number of solid state thermal decomposition processes show an obedience to the form required by classical thermodynamics. Many dehydration reactions illustrate this point.

First consider the general equation:

$$x\ R_s \rightleftharpoons yP_s + zP_g$$

The equilibrium constant (Kp) is given by;

$$Kp = P_g{}^Z$$

$$\text{or} \quad P_{pg}{}^Z = \text{constant}$$

where P_{pg} is the equilibrium pressure of P_g. It is the dissociation pressure at temperature T, (in degrees Kelvin). The standard free energy ($\Delta G°$) is given by

$$\Delta G° = -RT\ln Kp = -RT\ln P_{pg}$$

where R is the gas constant.

The relationship between $\Delta G°$ and the other standard thermodynamic functions is

$$\Delta G° = \Delta H° - T\Delta S°$$

where $\Delta H°$ and $\Delta S°$ represent the standard enthalpy of reaction and the standard entropy of reaction respectively. It follows that

$$\frac{d(\Delta G°)}{dT} = -\Delta S°$$

Substitution for $\Delta S°$ gives the Gibbs-Helmholtz equation

$$\Delta G° = \Delta H° + T\frac{d(\Delta G°)}{dT}$$

From $\Delta G° = -RT\ln Kp$ it also follows that

$$\frac{d(\Delta G°)}{dT} = -R\ln Kp - RT\frac{d\ln Kp}{dT}$$

Substitution into the Gibbs-Helmholtz equation gives

$$\Delta H° = RT^2\frac{d\ln Kp}{dT}$$

General integration gives

$$\ln Kp = -\frac{\Delta H}{RT} + \text{constant}$$

$$\text{or} \quad \ln P_{pg} = -\frac{\Delta H}{RT} + \text{constant}$$

Now reverting back to the dehydration reaction,

there may be various stages[28] as in the case of alum;

$$Al_2(SO_4)_3.16H_2O \rightarrow Al_2(SO_4)_3.14H_2O + 2H_2O$$

$$Al_2(SO_4)_3.14H_2O \rightarrow Al_2(SO_4)_3.12H_2O + 2H_2O$$

and $Al_2(SO_4)_3.12H_2O \rightarrow Al_2(SO_4)_3.9H_2O + 3H_2O$

Thus at any particular partial pressure of water the TG will show three stages (there might be more but the above three show clear steps on the TG plot). The real drawback to actually doing this kind of work on a TG balance is that most commercial units can only operate at one atmosphere. This kind of work requires specially constructed equipment capable of holding a good vacuum or any other partial pressure. A simple manometer system is probably best to obtain data on a single stage system but for multistage processes the TG balance operated at various pressures of product gas is probably the best technique to use, as the mass loss also allows the solid product phase to be identified.

Thus in oxide dissociations, it allows the possibility in a single run at a given partial pressure of oxygen to follow the multiple dissociation processes in the sequence;

$$MnO_2 \rightarrow Mn_2O_3 \rightarrow Mn_3O_4 \rightarrow MnO$$

It is of course possible to follow this on commercial instruments by using mixtures of oxygen and nitrogen at a total pressure of one atmosphere. However there is a real possibility that experiments in the presence of an inert gaseous species will not correspond to experiments in the presence of only the product gas oxygen at various pressures.

It is also possible to follow in the same way the decomposition of complex oxysalts, such as the dolomites or complex amine complexes.

6 KINETIC DATA

The classical method of following the kinetics of decomposition of a reaction is to operate the TG unit at a series of temperatures. In this isothermal method the thermal decomposition is followed by plotting the fraction decomposed (α) against time keeping the temperature constant and to repeat this experiment at 5 or 6 different temperatures. One can then identify the reaction mechanism and determine the specific reaction rate constant (k(T)) at each temperature. The Arrhenius parameters (the pre-exponential term (A) and the Activation Energy (E)) can then be calculated from a plot of log k(T) versus 1/T where T is the corresponding temperature of the isothermal experiment in Kelvin.

Initially kinetic models had to be established[29] and the mathematical relationships established between the fraction decomposed and the time of heat treatment. Most of these kinetic models for solid state decomposition are based upon the appearance of nuclei and the consequent result of their subsequent growth via a reaction interface. Usually the nucleation takes place at the particle surface, as a consequence of the experimental treatment. Subsequent growth then reflects the geometry of the contracting area of interface often imposed by the original shape and surface of the decomposing particles.[30] Later studies explored the effect of the additional process of diffusion of species away from or toward the reaction interface.[31] Specific mathematical relationships have become associated with particular models.[32] This is a wrong approach as the studies by Mampel,[33] Erofeev[34] and Avrami[35] show that the same model can give rise to various mathematical relationships. Nevertheless, it is advantageous to specify these relationships and to give them the accepted identifying code and this data is recorded in Table 1.

In gasification processes such as the oxidation of carbon blacks the reaction interface may be the surface area as measured in adsorption experiments. In other examples of thermal decomposition this identification of the reaction interface with the surface area is not permissible. The point made above regarding the same

Table 1   Broad   Classification   of   solid-state   rate
expressions

	$g(\alpha)=kt$	$f(\alpha)=1/k(d\alpha/dt)$
1) Acceleratory α - time curves.		
P1 Power law	$\alpha^{1/n}$	$n(\alpha)^{(n-1)/n}$
E1 Exponential law	$\ln \alpha$	α
2) Sigmoidal α - time curves.		
A2 Avrami-Erofeev	$[-\ln(1-\alpha)]^{1/2}$	$2(1-\alpha)(-\ln(1-\alpha))^{1/2}$
A3 " "	$[-\ln(1-\alpha)]^{1/3}$	$3(1-\alpha)(-\ln(1-\alpha))^{2/3}$
A4 " "	$[-\ln(1-\alpha)]^{1/4}$	$4(1-\alpha)(-\ln(1-\alpha))^{3/4}$
B1 Prout-Tompkins	$\ln[\alpha/(1-\alpha)] + c$	$\alpha(1-\alpha)$
3) Deceleratory α - time curves.		
3.1) based on geometrical models.		
R2 Contracting area	$1-(1-\alpha)^{1/2}$	$2(1-\alpha)^{1/2}$
R3 Contracting volume	$1-(1-\alpha)^{1/3}$	$3(1-\alpha)^{2/3}$
3.2) based on diffusion mechanisms.		
D1 one dimensional	α^2	$1/2\alpha$
D2 two dimensional	$(1-\alpha)\ln(1-\alpha)+\alpha$	$(-\ln(1-\alpha))^{-1}$
D3 three dimensional	$[1-(1-\alpha)^{1/3}]^2$	$3/2(1-\alpha)^{2/3}(1-(1-\alpha)^{1/3})^{-1}$
D4 Ginstling-Brounshtein	$(1-2\alpha/3)-(1-\alpha)^{2/3}$.	$3/2((1-\alpha)^{-1/3}-1)^{-1}$
3.3) based on "order" of reaction.		
F1 first order	$-\ln(1-\alpha)$	$1-\alpha$
F2 second order	$1/(1-\alpha)$	$(1-\alpha)^2$
F3 third order	$[1/(1-\alpha)]^2$	$0.5(1-\alpha)^3$

kinetic model leading to different kinetic relationships can be shown quite readily. In the simplest case the rate of decomposition per unit area of interface can be regarded as constant. Any variation in rate is then a consequence of variation in shape and size of the reaction interface area. Thus

$$\frac{d\alpha}{dt} = kR_s(t)$$

where α is the fraction decomposed, t is the time of heating, k is a proportionality constant and $R_s(t)$ is the area of the reaction interface. This latter term must also be considered a function of time. The decomposition of a plate-like crystal demonstrates the effect of particle shape on the reaction kinetics. On the assumption that there is a constant rate per unit area and that contributions from the edges are negligible then

$$\frac{d\alpha}{dt} = k \qquad \text{or } \alpha = kt$$

This kind of zero order reaction is shown by silver mellitate.[36] In the case of carbon blacks, however, it is the edge atoms which react many times faster than the basal plane atoms in oxidation by air.[37] Although this presents a different model the relationship is still zero order. Other kinds of zero order are based on different geometrical models, e.g. dehydration of cobalt oxalate.[38]

In the case of cylindrical shaped particles decomposing, the normal contracting area equation results,

$$kt = 1-(1-\alpha)^{\frac{1}{2}}$$

but if the particles are large then a first order relationship develops,

$$\frac{d\alpha}{dt} = k(1-\alpha)$$

The same is true for the contracting sphere model. In both these cases as the particle size diminishes the model predicts first a zero order, then a first order relationship and finally a contracting area or contracting volume equation. We thus have different models giving the same equation and the same model giving perhaps two or three alternative expressions. In such circumstances the analysis of the isothermal TG curve gives the mathematical relationship but the model can only be obtained by a considered judgement of data collected from a variety of techniques.

It can be argued that in solid state decompositions there is no such thing as an isothermal experiment. The reaction is going to be endothermic or exothermic and this fact should have an impact on the temperature of the reaction. This is a cogent reason for developing rising temperature kinetic experiments but is not the main reason. It is the thought that rising temperature methods could save time in obtaining kinetic analysis that promotes the real interest in this field. TG offers a convenient thermal analysis technique for solid state decomposition analysis but almost all established kinetic analysis techniques can easily be adapted to a rising temperature experiment. The difficulty is that most investigators prefer to use integral kinetic equations and in non-isothermal techniques one is faced with the integration of the function $\int e^{-E/RT} dT$.

A further restriction in the use of the rising temperature technique is the choice of the correct mechanistic kinetic equation. A usual method is to plot the Arrhenius relationship for each mechanistic equation in turn and choose the best straight line. This is only realistically possible with the aid of computer programs. It is easier to inspect the differential plot of the α–T data (Figure 3). Each type of mechanism gives a characteristic peak.[39] These can be tested exhaustively beforehand by use of computer programs which will plot the α–T curve if the mechanism and both A and E are entered into the program. Extensive use of idealized created curves of this kind enable the characteristic features of the DTG curve for any specified mechanistic equation to be characterized. The $d\alpha/dT$ against T plot is shown in Figure 3 together with appropriate features and labelling that are needed in the characterization. In this plot T_i represents the initial temperature or onset of the decomposition and T_f the final temperature, with

T_p representing the peak temperature. The vertical can be drawn from T_p to the base line and a horizontal line drawn at the half peak height. The half peak width can then be ascertained designated here as LoT (low temperature) and HiT (high temperature) with the half width HiT-LoT. As the curves may be asymmetric it is in fact better to use ΔLoT as the temperature range for half width on the LoT side with ΔHiT as the value on the HiT side. From these observations it is possible to identify the appropriate kinetic equation. Several schemes are possible.[39] The method shown here is presented in Figure 6 as a flow sheet. Each kinetic mechanistic equation is designated F_1, F_2, A_2 etc. as in Table 1 where the details of the kinetic behavior can be ascertained. The characteristic behavior features from the DTG plot (see Figure 3) are the value of α at which the rate of decomposition is maximum, the half width, and a note regarding the character of the initial temperature of decomposition (T_i) and the final temperature (T_f). Initial selection of the mechanistic kinetic equation for any particular TG plot is then based on the value of α_{max} (i.e. the value of α at the maximum rate of decomposition). Further separation is then based on the behavior of the onset and initial temperatures of the peak (T_i and T_f). If the onset is diffuse this is indicated $T_i(d)$ or if the onset is sharp as $T_i(s)$, and the same for T_f. A final separation from the DTG is possible based on the half width, noted on the flow sheet within parenthesis, e.g. (24-34) etc. This still leaves in one or two cases, the choice of two mechanisms to be resolved. It is suggested that these are resolved by resorting to the Arrhenius plot and noting which of the two gives the best linear response.

The calculation of the Arrhenius parameters from the TG plot is then relatively easy. Three basic equations are involved; the first is the linear temperature regime imposed on the sample;

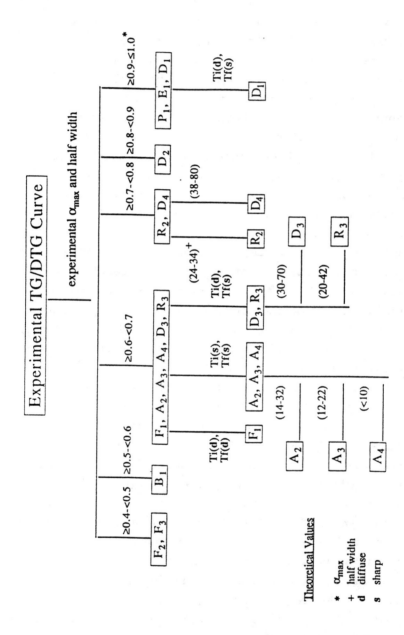

Figure 6 Flow chart showing procedures in recognizing the Kinetic Equations.

$$T = T_o + bt$$

where T is the temperature, T_o is the starting temperature (all in Kelvin), b is the heating rate, and t is the time of heating. The Arrhenius equation is the second equation

$$k = A \exp(-E/RT)$$

with the usual significance attached to all the terms and the differential form of the kinetic equation is used

$$\frac{d\alpha}{dt} = kf(\alpha)$$

The function $f(\alpha)$ has already been identified by the above procedures.

Noting that $b = \frac{dT}{dt}$ then $k = \frac{(d\alpha).b}{\frac{(dT)}{F(\alpha)}}$

when

$$\ln\left[\frac{\frac{(d\alpha).b}{(dT)}}{f(\alpha)}\right] = \ln A - \frac{E}{RT}$$

and the values of A and E can also be calculated.

7   REFERENCES

1.   K Honda, <u>Sci. Rep. Tohoku Univ.</u>, 1915, <u>4</u>, 97.
2.   K.Honda and H.Takagi, <u>Sci. Rep. Tohoku Univ.</u>, 1911, <u>1</u>, 229.
3.   J.Black, 'Experiments Upon Magnesum Alba, Quicklime and Other Alkaline Substances', Edinburgh, 1782.
4.   B.Higgins, 'Experiments and Observations Made with the View of Improving the Art of Composing and Applying Calcareous Cements and of Preparing Quick-lime: Theory of These Arts: and Specifications of the Authors Cheap and Durable Cement for Building, Incrustation or Stuccoing, and Artificial Stone', London, 1780.
5.   See J.R.Partington, 'Chemical Thermodynamics', 3rd Ed., Constable, Edinburgh, 1940, p.32.
6.   T.J.Seebeck, <u>Annln. Phys.</u>, 1826, <u>6</u>, 130, 253.
7.   A.C.Becquerel, <u>Ann. Chem.Phys.</u>, 1826, <u>31</u>, 371; 1863, <u>68</u>, 49.
8.   C.S.M.Pouillet, <u>Compt. Rend.</u> Paris, 1836, <u>3</u>, 782.
9.   H.L. Chatelier, <u>Compt. Rend.</u>, Paris, 1886, <u>102</u>, 819.
10.   C.W.Siemens, <u>Proc. Roy. Soc.</u>, 1871, <u>19</u>, 443.
11.   H.L.Chatelier, <u>Compt. Rend.</u>, Paris, 1892, <u>11</u>, 214.
12.   'Report of a Symposium on Microbalances', 22 Sept. 1949, published by The Royal Institute of Chemistry.

13. E.Warburg and T.Ihmori, <u>Ann. Phys. Chem. NF</u>, 1886, <u>27</u>, 481.
14. W.Nernst and E.H.Riesenfeld, <u>Chem. Ber.</u>, 1903, <u>36</u>, 2086.
15. E.Steele and K.Grant, <u>Proc. Roy. Soc.</u>, 1909, <u>A82</u>, 580.
16. R.W.Gray and Sir W.Ramsay, <u>Proc. Roy. Soc.</u>, 1912, <u>A86</u>,278.
17. O.Brill, <u>Z. Anorg. Allg. Chem.</u>, 1905, <u>45</u>, 272.
18. A.R.McGhie, <u>Anal. Chem.</u>, 1983, <u>55</u>, 987.
19. S.D.Norem, M.J.O'Neill and A.P.Gray, <u>Thermochim. Acta</u>, 1970, <u>1</u>, 29.
20. T.A.Evans, Ph.D. thesis, Univ. of Toledo, 1991.
21. D.Dollimore and D.L.Griffiths, <u>J. Thermal Analysis</u>, 1970, <u>2</u>, 229.
22. C.N.Fenner, <u>Am. J. Sci.</u>, 1913, <u>36</u>, 331.
23. P.Davies, D.Dollimore and G.R.Heal, <u>J.Thermal Analysis</u>, 1978, <u>13</u>, 473.
24. D.Dollimore and J.Pearce, <u>J. Thermal Analysis</u>, 1974, <u>6</u>, 321.
25. D.Dollimore, <u>Thermochim. Acta</u>, 1980, <u>38</u>, 1.
26. D.Dollimore, D.L.Griffiths and D.Nicholson, <u>J. Chem. Soc.</u>, <u>1963</u>, 2617.
27. J.Azizi, D.Dollimore, C.C.Philip, W.A.Kneller and P. Manley, 'Proc. 18th North Amer. Thermal Analysis Soc.', (Editor I.R. Harrison), 1989, p.980.
28. P.Barret and R.Theard, <u>Compt. Rend. Seance Acad. Sci.</u>, Paris, 1965, <u>260</u>, 2823.
29. P.W.M.Jacobs and F.C.Tompkins in 'Chemistry of the Solid State' (Editor W.E.Garner) Butterworths, London, 1955, p.184.
30. C.J.Keattch and D.Dollimore, 'An Introduction to Thermogravimetry', 2nd Ed., Heyden, London, (1975), p.57.
31. S.F.Hulbert, <u>J. Brit. Ceram. Soc.</u>, 1968, <u>6</u>, 11.
32. J.Sestak, V.Satava and W.W.Wendlandt, <u>Thermochim. Acta</u>, 1973, <u>7</u>, 333.
33. K.L.Mampel, <u>Z. Phys. Chem. Abt.</u>, 1940, <u>A 187</u>, 43, 235.
34. B.V.Erofeev, <u>C.R. Dokl, Acad. Sci. USSR</u>, 1946, <u>52</u>, 511.
35. M.Avrami, <u>J. Chem. Phys.</u>, 1939, <u>7</u>, 1103; 1940, <u>8</u>, 212; 1941, <u>9</u>, 177.
36. R.J.Acheson and A.K.Galwey, <u>J. Chem. Soc.A</u>, 1968, 942.
37. J.Azizi, D.Dollimore, C.C.Philip, W.A.Kneller and P. Manley in 'Proc. 18th North American Thermal Analysis Society' (Editor I.R. Harrison), 1989, p.980.
38. D.Broadbent, D.Dollimore and J.Dollimore, <u>J.Chem. Soc.A</u>, 1966, 1491.
39. D.Dollimore, T.A.Evans, Y.F.Lee and F.W.Wilburn in 'Proc. 19th North Amer. Thermal Analysis Society' (Editor I.R. Harrison), 1990, 397.

Complementary Thermal Analysis Techniques

E. L. Charsley

THERMAL ANALYSIS CONSULTANCY SERVICE, LEEDS METROPOLITAN UNIVERSITY,
CALVERLEY STREET, LEEDS LS1 3HE, UK

1 INTRODUCTION

It is now generally recognised that TG, DTA or DSC alone
cannot be used for a complete interpretation of the
reactions of a given system and need to be supplemented
by other thermal methods and general analytical
techniques. In this chapter some of the thermal
techniques that can be used will be considered and the
advantages of carrying out measurements simultaneously,
i.e. on the same sample at the same time, will be
discussed.[1] The latter approach eliminates the
uncertainty inherent in the comparison of results
obtained using two or more individual thermal analysis
units. This is particularly valuable in the case of
complex decomposition reactions.

2 SIMULTANEOUS TG-DTA AND TG-DSC

Comparison between TG and DTA/DSC

Simultaneous TG-DTA and TG-DSC are the most widely
used of the simultaneous techniques due to their
complementary nature. TG is a technique that, although
limited in scope to those reactions taking place with a
change in weight, gives results that are intrinsically
quantitative. Thus once the balance has been calibrated,
and any allowance made for "buoyancy effects", the
measured weight losses will faithfully reflect the

overall reaction taking place. The results will not be influenced by experimental factors, such as crucible material or atmosphere, unless these influence the reaction directly. A stable baseline will be given when a reaction is not taking place and this baseline will not be altered by variations in the programmed heating rate or by physical changes in the sample such as melting or sintering or by changes in specific heat capacity.

In contrast, DSC and DTA are considerably more versatile techniques and can detect any reaction which takes place with a change in energy. However, careful calibration using heat capacity measurements and/or materials with known heats of transition is necessary to obtain quantitative results from DSC measurements and quantitative measurements are only possible with certain types of DTA equipment. The results obtained from DSC or DTA are dependent on the experimental conditions and since both the nature of the atmosphere and the crucible type and material affect the results directly, calibration must be carried out under the same conditions as the measurements are to be made. The instrument baseline will reflect changes in the heat capacity of the sample and will also be influenced by changes in the disposition of the sample in the crucible due to e.g. melting, bubbling or sintering. These factors can sometimes lead to complications in the interpretation of the results, particularly at high temperatures.

The techniques are therefore complementary in nature and the TG curve can often be used to aid in the interpretation of DTA and DSC data. Some objections were initially made to simultaneous TG-DTA on the grounds that the results were a compromise in that neither technique could be run under its optimum experimental conditions.[2] Thus it was suggested that TG experiments needed large samples and slow heating rates while DTA required smaller samples and higher heating rates. These requirements can be attributed primarily to design limitations in early thermal analysis equipment. Thus TG experiments used large samples due to low balance sensitivities and hence required low heating rates for good resolution. Sample size in DTA experiments was limited by space considerations within the sample block and hence fast heating rates were used to obtain suitable signals at the relatively low amplification levels then available.

It is clearly desirable that the conditions chosen for a thermal analysis experiment should be governed by the system being studied, rather than by the instrument being used and hence the conditions should be similar for both TG and DTA or DSC studies. With modern thermobalances it is possible to use very small sample weights and hence obtain good resolution at heating rates of $10^{\circ}C$ min^{-1} or more. Modern DTA and especially DSC equipment is sufficiently sensitive to allow slow heating rates to be used where required and therefore there is considerable overlap in the performance capabilities of the two techniques. This enables high quality simultaneous TG-DTA and TG-DSC measurements to be made for a wide range of materials.

Although in many cases results obtained from different TG and DTA/DSC instruments can be compared, there are a number of factors which make it advantageous to carry out the measurements simultaneously. These include differences in the sample heating rates due to different thermal lags in individual instruments, differences in self-heating/cooling of the sample due to different thermal environments and uncertainties in the sample temperature in the TG measurements. More importantly, although the same gas flow rate may be used in TG and DTA/DSC experiments it is unlikely to have the same purging effect on the sample in different types of equipment and hence may lead to significant differences in the curves for decomposition or gas-solid reactions obtained separately.

For complex reactions, where there are several overlapping stages, it may be difficult to match the energy changes and weight losses when the reactions are carried out separately. This is particularly true when decomposition takes place from a bubbling melt, where there may in any case be small variations from experiment to experiment.

Apparatus

The normal method of constructing a TG-DTA or TG-DSC apparatus is to incorporate a DTA or DSC head into a thermobalance. Probably the earliest examples date from 1957 and Paulik and Paulik have reviewed the development of the technique.[1] Much pioneering work was carried by the Pauliks in Hungary using the Derivatograph[3] and by Wiedemann in Switzerland using the Mettler TA1.[4] Today, although not as widely available as separate TG and DTA or DSC equipment, there is a good choice of simultaneous equipment from a number of manufacturers. The most important recent development is the introduction of heat-flux DSC heads enabling simultaneous TG-DSC measurements to be carried out.

The main problem to be overcome in the construction of TG-DTA or TG-DSC equipment is to obtain the thermocouple output from the DTA or DSC head without affecting the action of the balance. This is normally carried out by using fine wires or ribbons, made of the same material as the thermocouples, to make the connection from the head to the measuring circuit.

An example of a simultaneous TG-DSC equipment (Netzsch Model STA 409) is shown in Figure 1.[5] The instrument features a platinum v platinum-10% rhodium heat-flux DSC plate-type head, mounted on a top-loading electronic microbalance. The platinum-rhodium wound furnace, specifically designed for DSC studies, gives an operating range of ambient to 1500°C. An example of a quantitative TG-DTA unit[6] operating over the same temperature range (Stanton Redcroft Model STA 1500) is

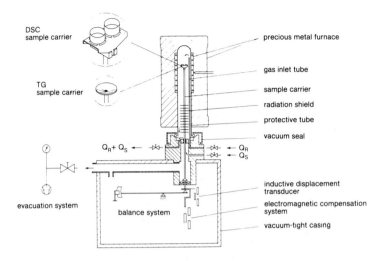

Figure 1 Cross-section of Netzsch Model STA 409 simultaneous TG-DSC apparatus

described later in this book in the chapter on evolved gas analysis.

The Setaram Model TG-DSC 111[7] takes advantage of its Calvet type DSC cell, where the sample is not in direct contact with the detector, which is in the form of a cylindrical thermopile. This enables the sample to be suspended directly in the thermopile detector from an electronic microbalance as shown in Figure 2. The reference crucible is suspended in a similar thermopile and the resulting symmetrical arrangement reduces the "buoyancy effect" on the TG curve.

Instruments are also available for sub-ambient experiments. Thus an instrument has been described which uses a liquid nitrogen cooled furnace to enable TG-DSC measurements to be carried out over the range -125°C to 625°C.[8]

Sample Temperature Measurement

One of the problems in carrying out TG experiments in the majority of commercial equipment, is the

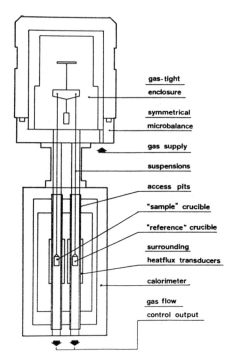

Figure   2   Cross-section   of   Setaram   Model   111
simultaneous TG-DSC unit

uncertainty   in   the   sample   temperature   since   the
measuring thermocouple is not in direct contact with the
sample crucible. TG data obtained using simultaneous
equipment does not suffer from this problem since the
sample temperature is measured directly. Also since the
DTA   or   DSC   signal   is   measured   simultaneously,   the
temperature signal of the instrument can be calibrated
using the melting points of pure metals or the Certified
Reference   Materials   developed   by   the   International
Confederation   for   Thermal   Analysis.[9]   The   need   for
calibration using magnetic transitions[10] or the drop-
weight   method   using   fusible   links[11]   is   therefore
avoided. Indeed TG-DTA or DSC can be used to determine
more   accurate   values   for   the   magnetic   reference
materials used in the calibration of thermobalances.[9]

Thermal Stability

An example of the use of simultaneous TG-DTA for the study of thermal stability is the fusion of potassium chloride, which was being investigated in the author's laboratory as a possible enthalpy standard for the calibration of DTA equipment. Samples which had been removed from a DTA apparatus after cooling were found to have lost a small amount of weight. The sample was investigated by TG-DTA to see if the weight loss occurred after the fusion and hence would not influence the enthalpy measurements. The curves are given in Figure 3 and show that the sample lost weight by

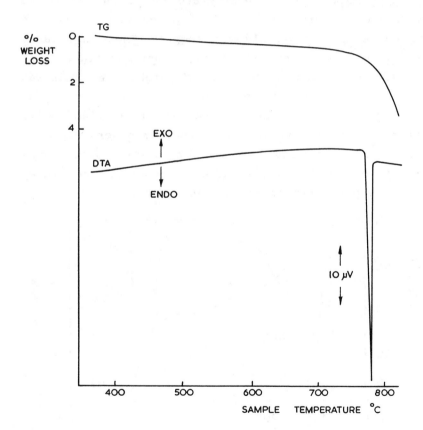

Figure 3 TG-DTA curves for potassium chloride (sample weight:10mg,heating rate:10°Cmin⁻¹,atmosphere:nitrogen)

sublimation before fusion and that the rate of weight loss increased after fusion. This information, which revealed the need to use a crucible lid to reduce the weight loss, is difficult to obtain by indirect weighing. A similar investigation was carried out on the thermal stability of the organic materials, formerly marketed by the National Physical Laboratory as enthalpy standards for the calibration of DSC equipment.[12]

Decomposition Reactions

The ability to correlate energy and weight changes in complex reactions is illustrated by the decomposition of magnesium nitrate hexahydrate shown in Figure 4. While the TG curve shows a three stage weight loss the DSC curve is considerably more complex. There is a large peak in the region of 90°C which was not associated with any discrete change in the rate of weight loss. This was shown by thermomicroscopy[13] (see Section 3) to be due to an aqueous fusion caused by the salt dissolving in its water of crystallisation. The shoulder observed before this reaction was found to be due to a solid-solid phase transition which could be clearly observed when water loss was prevented by using a sealed DSC pan.

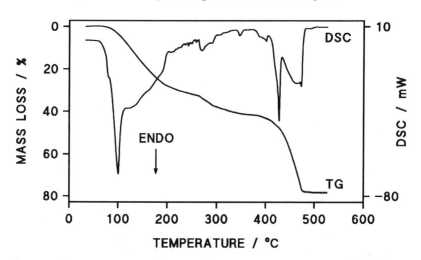

Figure 4 TG-DSC curve for magnesium nitrate hexahydrate (sample weight:10mg, heating rate:10°Cmin⁻¹,atmosphere: nitrogen)

In the region of 400°C a second sharp endothermic reaction was given, again without any weight loss step. Direct observation showed that this was due to the fusion of the partially decomposed anhydrous magnesium nitrate formed in the dehydration reaction.

TG as an Aid in the Interpretation of DTA or DSC Curves

As mentioned earlier the TG curve will show a stable baseline when there is no reaction taking place and can be used to help define the DTA or DSC baseline. This is illustrated in Figure 5 by the TG-DSC curve for the decomposition of an uncured polyimide resin.[8] Following glass transitions and the fusion of the resin in the region of 120°C, a curing exotherm was given with a peak temperature of about 280°C. The TG trace showed that the initial portion of the curing reaction was

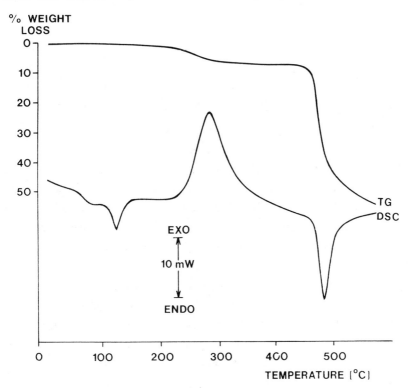

Figure 5 TG-DSC curve for a polyimide resin (sample weight:10mg, heating rate:10°Cmin^{-1}, atmosphere: nitrogen)

accompanied by a weight loss of about 7% due to loss of volatiles. The TG curve showed that the endothermic decomposition of the cured resin did not begin until about 450°C and this enabled the baseline for the curing reaction to be established more accurately.

The ability to measure the exact weight loss at any stage during the DTA or DSC peak offers a number of important advantages. Thus Henderson and Emmerich[14] were able to use this information to determine the specific heat of a glass-filled polymer composite above its decomposition temperature, while Etzler and Conners have proposed a simultaneous TG-DSC method to determine heats of vaporisation.[15] It would also be possible to validate the suitability of using a DSC or DTA curve for kinetic measurements on a given reaction involving a weight loss, by direct comparison with the TG curve.

3   THERMOMICROSCOPY

Introduction

The ability to observe a sample as it is heated under conditions of controlled atmosphere and heating rate can provide a valuable supplement to other thermal analysis techniques. To take a simple example, although the technique of DSC has reached a high level of sophistication, it is non-specific and cannot distinguish between a phase change and a fusion reaction. This is readily accomplished by thermomicroscopy and using this technique it is possible to observe directly such phenomena as phase changes, fusion, decomposition reactions and processes such as sintering, decrepitation, creeping of a liquid melt and foaming or bubbling reactions which can often complicate the interpretation of a DSC or DTA curve.

Two modes of operation will be considered in this section. In the first mode the sample is viewed by reflected light and in the second the sample is observed by means of transmitted light, normally under polarizing conditions.

Reflected Light Thermomicroscopy

An example of a hot stage unit[13] for observation of a sample under reflected light conditions over the temperature range ambient to 1000°C is given in Figure 6. The system enables the sample to be heated in a standard DSC pan, under a controlled atmosphere, while being observed using a stereo microscope. Incorporation of a photocell into the microscope head enables the intensity of the light reflected from the surface of the sample to be recorded as a function of temperature. This considerably reduces the problems in assessing visually gradual changes in the colour of samples and enables the temperature of rapid reactions, such as fusion, to be recorded.

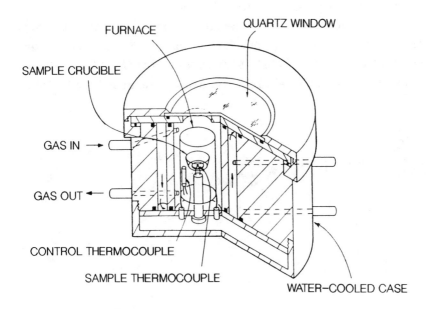

Figure 6 Cross-section of Stanton Redcroft HSM5 hot stage unit for reflected light measurements

An example of the reflected light technique is illustrated in Figure 7 which shows the reflected light curve for magnesium nitrate hexahydrate.[13] As discussed earlier, thermomicroscopy was used to identify the aqueous fusion of the sample in the region of 90°C and also the fusion of the partially decomposed magnesium nitrate in the region of 410°C. These reactions are clearly recorded on the reflected light curve together with the reduction in light intensity above 130°C as the liquid becomes opaque and viscous. The noise on the trace caused by the bubbling reaction can also be observed.

A similar approach was used by Haines and Skinner who fitted a stereo microscope and photo-detection system to a Perkin-Elmer DSC to enable simultaneous DSC and reflected light intensity measurements to be carried out.[16,17]

Transmitted Light Thermomicroscopy

Although very valuable complementary thermal information can be observed under reflected light conditions, additional structural information may be obtained when the sample is viewed in transmitted light.

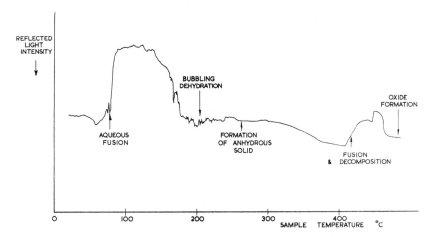

Figure 7 Reflected light intensity curve for magnesium nitrate hexahydrate (sample weight:8mg, heating rate:10°Cmin⁻¹, atmosphere:argon)

The application of transmitted light thermal microscopy in the field of organic chemistry is well documented in the work of the Koflers[18] and of McCrone[19]. The value of the method, particularly in the polymer field, was extended by adapting the apparatus to carry out depolarized light intensity measurements. This technique, first applied to polymer systems by Magill[20], detects transitions in birefringent materials by measuring the intensity of the polarized light transmitted by the sample as a function of temperature.

An example of a transmitted light hot stage unit (Linkam Model TH 600)[21], designed to operate over a much wider temperature range than the conventional Kofler units under conditions of controlled heating rate and atmosphere, is shown in Figure 8. The sample is heated by means of a silver block containing a nichrome heating element and fitted with a sapphire or quartz window to allow the sample to be viewed by transmitted light. The integral liquid nitrogen cooling system enables controlled operation of the system to be carried out over the temperature range -180 to 600°C. The unit was used in conjunction with a polarizing microscope fitted with a silicon photo-detector for the depolarised light intensity measurements. For these measurements the sample is viewed under crossed polars so that the only light transmitted is due to rotation of the polarized light by the crystalline structure of the material.

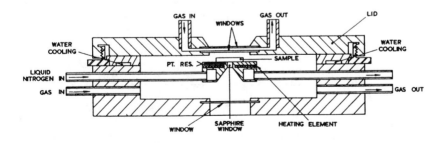

Figure 8   Cross-section of Linkam TH 600 transmitted light hot stage unit

Thus any change in the structure of the material on heating will result in changes in the recorded light intensity. Melting of the sample will result in extinction of the light. An example of a depolarized light intensity curve is given in Figure 9, which shows the complex behaviour observed on heating a liquid crystal sample.[22] The response of the optical system is extremely rapid and will provide a much faster response than a DSC system. This has recently been discussed by Eysel and co-workers who compared the DSC and thermomicroscopy behaviour of a number of inorganic single crystals.[23]

Commercial equipment is also available which enables DSC and thermomicroscopy measurements to be carried out simultaneously.[24,25] The system (Mettler Model FP84) which is shown in Figure 10, takes advantage of a DSC sensor in the form of a thin film thermopile

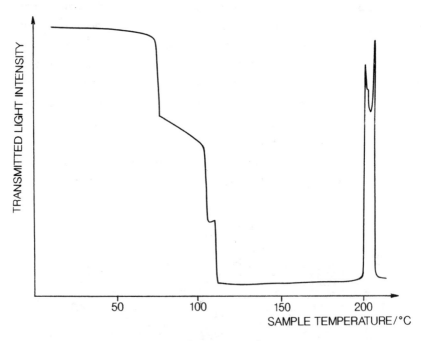

Figure 9 Depolarized light intensity curve for a liquid crystal sample (B.D.H. Type T27), showing crystalline → smectic G → smectic E → smectic A → nematic → isotropic transitions (heating rate:10°C min^{-1}, atmosphere: nitrogen)

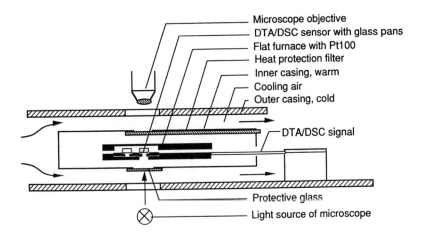

Microscope objective
DTA/DSC sensor with glass pans
Flat furnace with Pt100
Heat protection filter
Inner casing, warm
Cooling air
Outer casing, cold

DTA/DSC signal

Protective glass
Light source of microscope

Figure 10 Cross-section of Mettler FP 84 DSC -thermomicroscopy unit

mounted on a glass disc to enable the sample to be viewed by transmitted light while simultaneously recording a DSC curve. Heating is carried out by means of two metal plates with embedded heating elements, providing a maximum temperature of 300°C.

An example of the use of the unit is in the study of thermal behaviour of caffeine.[24] The DSC trace is shown in Figure 11. The peak at 141°C corresponds to the first order β to α transition and the second to melting at 236.5°C. The drift in the baseline was found to be due to sublimation of the sample and vaporisation of the sample during melting gave rise to the bubbles shown in the accompanying photomicrographs.

The illumination for thermomicroscopy studies is not restricted to tungsten light. Thus Kagemoto et al have discussed recently the development of a DTA apparatus equipped with a laser to study simultaneously the thermal and optical properties of biopolymer solutions.[26]

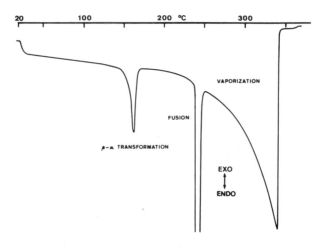

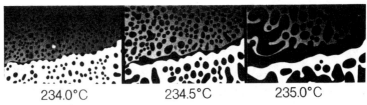

234.0°C   234.5°C   235.0°C

Figure 11 DSC curves and photomicrographs for caffeine
(sample weight:9.8mg, heating rate:5°C min^{-1}, atmosphere:
air) (From Ref.24)

4   THERMAL   ANALYSIS-HIGH   TEMPERATURE   X-RAY
DIFFRACTION

X-ray diffraction (XRD) methods are used extensively to
identify phases formed during thermal analysis studies.
Measurements are normally performed on samples which
have been removed from the equipment at different
temperatures during the thermal analysis experiment.
This approach can suffer from a number of limitations
such as the inability to study high temperature forms
that undergo a reversible reaction on cooling and the
amount of time taken to generate accurate reaction
profiles.

It is therefore advantageous to carry out XRD
measurements simultaneously with the thermal analysis
experiments. Although comparatively few simultaneous
instruments have been developed, due to the experimental

difficulties involved, an example of the power of the simultaneous XRD technique is taken from a review by Barret.[27] He used the apparatus developed by Gerard[28] to enable simultaneous TG-XRD measurements to be carried out.

A diagram of the equipment is shown in Figure 12.

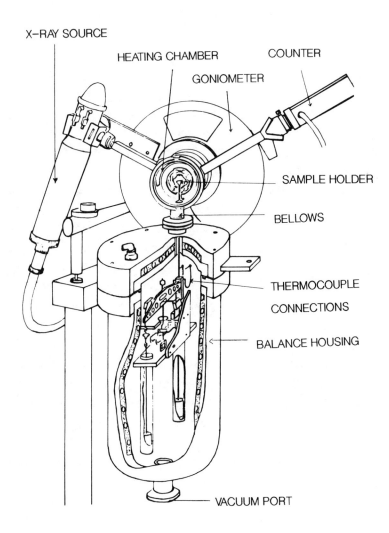

X-RAY SOURCE

HEATING CHAMBER

COUNTER

GONIOMETER

SAMPLE HOLDER

BELLOWS

THERMOCOUPLE

CONNECTIONS

BALANCE HOUSING

VACUUM PORT

Figure 12 Diagram of apparatus for simultaneous TG-XRD measurements (from Ref.27)

The sample is spread in a thin layer on a plate supported on an alumina rod which also carries the sample thermocouple. A single layer of sample was used, so that the diameter of the crystals did not exceed the penetration depth of the X-ray diffraction beam, in order to have a reasonable correlation between the XRD and TG measurements.

Barret[27] has discussed the application of this instrument to the study of the reduction of tungsten oxide (WO_3) to tungsten by heating in hydrogen. Under isothermal conditions the TG curves suggested an apparently single stage reduction process (Figure 13a). However use of the simultaneous XRD technique revealed that the reaction was considerably more complex and that $W_{20}O_{58}$, WO_2 and β tungsten were formed as reaction intermediates (Figure 13b).

Wiedemann and Bayer[29] described an alternative approach to simultaneous TG-XRD where a high temperature X-ray film camera was linked to a thermobalance. In this case geometrical and focusing problems involved in coupling the two instruments were avoided by following the weight changes by measuring the gaseous reaction products. The latter were passed to the thermobalance from the camera via a vacuum-tight link and formed a molecular beam which was directed to the balance pan. The force of the impacting molecules was then used to derive the weight change and the gases were also analysed using a quadrupole mass spectrometer. The system was used to study lunar dust samples.

The development of position-sensitive detectors which enable the relatively fast acquisition of data during heating experiments has provided an impetus to time- resolved high temperature X-ray diffraction techniques and an increasing number of papers are appearing in the literature.[30-33] In general, samples are heated in commercial high temperature X-ray diffractometer attachments at low scan rates eg 0.1-1°C min^{-1}. An example of this type of approach is given by

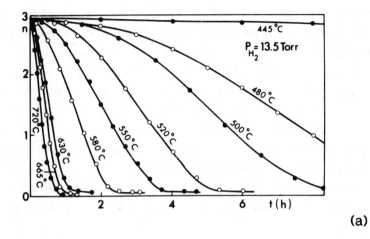

(a)

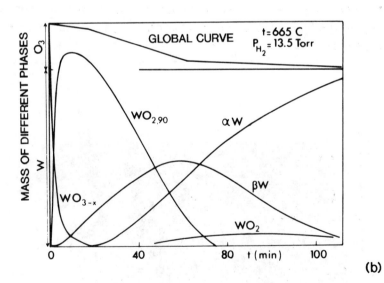

(b)

Figure 13   TG-XRD studies of the reduction of tungstic oxide in hydrogen; a) TG curves, b) XRD curves (from Ref.27)

Auffredric et al., who have used the technique to study
the decomposition of cadmium hydroxide nitrate
$Cd_3(OH)_5NO_3$.[31] The time-resolved XRD plots for an
experiment on the salt heated under vacuum conditions
are shown in Figure 14. The curves indicated that the
decomposition to CdO occurred in a single stage, the
intermediate $Cd(OH)NO_3$, observed in heating in nitrogen,
decomposing as it was formed. Plotting the integrated
intensities of selected diffraction lines as a function
of temperature, together with TG data obtained
separately, indicated that the rate of desorption of
gaseous reaction products was much lower than the rate
of structural transformation. This was attributed to the
adsorption of NO, NO_2 and O_2 on the finely divided CdO
surface.

Eisenreich and Engel[30] in their study of the
reaction of ammonium nitrate and copper oxide by time-
resolved high temperature X-ray diffraction have
discussed the use of the technique for kinetic studies
using a method of summing difference patterns from the
diffraction traces.

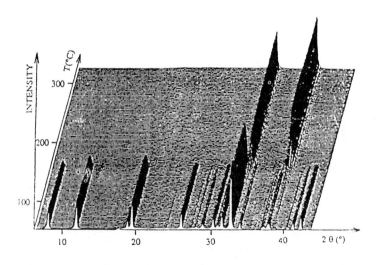

Figure 14   Time resolved XRD plots for $Cd_3(OH)_5NO_3$
(heating rate:5°C hour^{-1}, atmosphere:nitrogen) (from
Ref.31)

The advantages of carrying out DSC measurements in conjunction with time-resolved XRD studies has been realised. Thus Fawcett et al have described the applications of a simultaneous DSC-XRD system for operation over the range ambient to 450°C.[34] An example of the use of the equipment was in the study of the fusion of polyethylene. Samples were heated at 1.25°C min[-1] and XRD scans were taken over the 12-32° 2θ range for 4 minutes at 5 minute intervals.

Figure 15a shows every second XRD scan and the DSC curve obtained during melting. Fusion was detected by XRD as low as 70°C compared with about 100°C from the DSC trace. This could have resulted in a low result for the degree of crystallinity if measured by the latter method. Similarly on cooling (Figure 15b), the XRD scans showed that solidification was still occurring at 80°C and that this was not readily apparent from the DSC curve.

More recently the range of the instrument has been extended to 600°C and the versatility of the system increased by incorporation of an evolved gas analysis facility using a quadrupole mass spectrometer.[35]

The compact size of thermomicroscopy equipment has resulted in the use of hot stages for XRD measurements. Thus Russell and Koberstein have used the Mettler FP84 DSC-hot stage unit described earlier to carry out DSC-XRD measurements on polymers.[36] Similarly Caffrey has adapted the Mettler unit for DSC-XRD measurements on biological liquid crystals.[37] In the latter work, advantage was taken of the very high flux of synchrotron radiation to reduce considerably the exposure times.

5 CONCLUSIONS

Simultaneous TG-DTA and particularly TG-DSC will play an increasing role in the interpretation of complex decomposition mechanisms, particularly those taking place at high temperatures. In addition to the ability

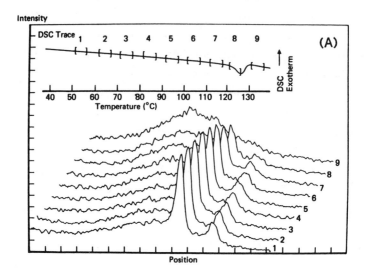

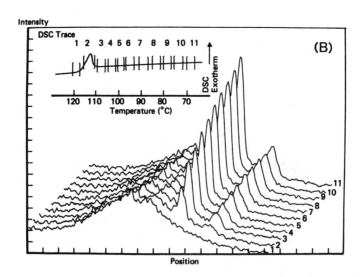

Figure 15 DSC-XRD curves for polyethylene; a) heating,
b)cooling (programme rate 1.25°C min⁻¹) (from Ref.34)

to provide a direct correlation of data from the two techniques, the facility of accurate temperature measurements for TG studies and the ability to use the TG data to aid the interpretation of DSC/DTA curves are powerful attractions.

Thermomicroscopy and high temperature X-ray diffraction measurements are also likely to see increasing use. For these techniques DSC or DTA is the most readily incorporated additional thermal method. In the absence of commercially available TA-XRD equipment, DSC-thermomicroscopy units will continue to provide a good basis for the construction of DSC-XRD units

Where the reactions being studied involve the production of gaseous products, evolved gas analysis techniques, particularly mass spectrometry and Fourier transform infrared spectroscopy, may be added to the simultaneous techniques described above. In this way a considerable amount of additional information may be obtained with only minor experimental modifications.

ACKNOWLEDGEMENTS

I would like to thank the publishers and instrument manufacturers who have given permission for the reproduction of diagrams.

REFERENCES

1. F.Paulik & J.Paulik, Analyst, 1978, 103, 417.
2. P.D.Garn, 'Thermoanalytical Methods of Investigation', Academic Press, 1965.
3. F.Paulik, J.Paulik & L.Erdey, Z. Analyt. Chem., 1958, 160, 211.
4. H.G.Wiedemann, Chemie Ing. Tech., 1964, 36, 1105.
5. W.D.Emmerich, E.Kaiserburgher, J.E.Kelly, E.Wassmer, Proc. 15th North American Thermal Society Conference, Cincinatti, 1986, 18.
6. E.L.Charsley, J.Joannou, A.C.F. Kamp, M.R.Ottaway, & J.P.Redfern, 'Thermal Analysis', Vol.1, Ed. H.G.Wiedemann, Birkhauser Verlag, 1980, 237.
7. P.Le Parlouer, Thermochim. Acta, 1985, 92, 371.
8. A.J.Brammer, E.L.Charsley, S.J.Marshall & J.Swan, Proceedings Thirteenth North American Thermal Society Conference,
9. E.L.Charsley, S.St.J.,Warne & S.B.Warrington,

Thermochim. Acta, 1987, 114, 53.

10. P.D.Garn,O.Menis & H.G.Wiedemann, J. Therm. Anal., 1981, 20, 185.

11. R.G.Blaine & P.G.Fair, Thermochim. Acta, 1983, 67, 233

12. P.A.Barnes,   E.L.Charsley,   J.A.Rumsey   & S.B.Warrington, Analyt. Proc. 1984, 21, 5.

13. E.L.Charsley &   A.C.F.Kamp, 'Thermal Analysis', Vol.1, Ed. H.G.Wiedemann, Birkhauser Verlag, 1972, 499.

14. J.B.Henderson & W.D.Emmerich, Proc. 16th North American Thermal Analysis Society Conference, Washington D.C., 1987, 215.

15. F.M.Etzler & J.M.Conners, Thermochim.Acta, 1991, 189, 155.

16. P.J.Haines & G.A.Skinner, Thermochim. Acta, 1982, 59, 343.

17. P.J.Haines & G.A.Skinner, Thermochim. Acta, 1988, 134, 201

18. L.Kofler & A.Kofler, 'Thermo-Mikromethoden', Verlag Wagner, 1954

19. W.McCrone, 'Fusion Methods in Chemical Microscopy, Intersience, 1957.

20. J.H.Magill, Polymer, 1961, [2], 221.

21. E.L.Charsley, A.C.F.Kamp &   J.A.Rumsey, 'Thermal Analysis', Vol.1, Ed. H.G.Wiedemann, Birkhauser Verlag, 1980, 285.

22. E.L.Charsley & A.C.F.Kamp, Poster presented at 9th ICTA Congress, Jerusalem, 1988

23. S.Hess & W.Eysel, J.Therm. Anal., 1989, 35, 627.

24. W.Perron,   G.Bayer   &   H.G.Wiedemann,   'Thermal Analysis' Vol.1, Ed. H.G.Wiedemann, Birkhauser Verlag, 1980, 279.

25. H.G.Wiedemann & G.Bayer, Thermochim. Acta, 1985, 92, 371.

26. A.Kagemoto, Y.Okado, H.Iree, M.Oka & Y.Baba, Thermochim. Acta., 1991, 176, 1.

27. P.Barret,, 'Vacuum Microbalance Technology', Ed. C.Eyraud & M.Escoubes, Vol.3, Heyden, 1975, 205.

28. N.Gerard, J. Physics, E, 1974, 7, 509.

29. H.G.Wiedemann & G.Bayer, Z. Anal. Chem., 1973, 266, 97.

30. N.Eisenreich & W.Engel, J.Therm, Anal., 1985, 35, 577.

31. J.P.Auffredic, J.Plevert & D.Louer, J. Therm. Anal., 1991, 37, 1727.

32. W. Engel, N. Eisenreich, Proceedings of the 5th European Symposium on Thermal Analysis & Calorimetry, Nice, France, 1991, J.Therm. Anal., in Press.

33. M. Epple & H.K. Cammenga, Proceedings of the 5th European Symposium on Thermal Analysis & Calorimetry, Nice, France, 1991, J.Therm. Anal. in Press.

34. T.G.Fawcett, C.E.Crowther, L.F.Whiting, J.C.Tou, W.F.Scott, R.A.Newman, W.C.Harris, F.J.Knoll &

V.J.Caldecourt, <u>Advances in X-Ray Analysis</u>, 1985, <u>28</u>, 227.

35. R.A.Newman, J.A.Blazy, T.G.Fawcett, L.F.Whiting & R.Stowe, <u>Advances in X-Ray Analysis</u>, 1987, <u>30</u>, 493

36. T.P.Russell & J.T.Koberstein, <u>J.Polym. Sci</u>, 1985, <u>23</u>, 1109

37. M.Caffrey, <u>Trends in Analyt. Chem.</u>, 1991, <u>10</u>,156.

Evolved Gas Analysis

S. B. Warrington

THERMAL ANALYSIS CONSULTANCY SERVICE, LEEDS METROPOLITAN UNIVERSITY,
CALVERLEY STREET, LEEDS LS1 3HE, UK

1 INTRODUCTION

Evolved gas analysis has been defined by the ICTA Nomenclature Committee[1] as: "a technique in which the nature and/or amount of volatile product(s) released by a substance is/are measured as a function of temperature as the substance is subjected to a controlled temprature programme." The recommended abbreviation is EGA.

The definition is a broad one, and encompasses other techniques which are practised under their own, sometimes unofficial titles. Thus emanation thermal analysis (ETA), temperature-programmed desorption (TPD) and temperature- programmed reduction (TPR) certainly fall within the scope of EGA, but will not be described here.

EGA is an improvement on the simpler technique of evolved gas detection (EGD), which merely detects, non-specifically and non-quantitatively, the evolution of a volatile product. EGD has been especially useful in conjunction with DTA, in deciding whether thermal events were accompanied by mass loss.[2]

The definition does not require EGA to be associated with any other type of measurement, and many valuable studies have been conducted by EGA alone. Indeed several of the familiar rate expressions describing solid-state decompositions grew from work in which the progress of the reaction was followed manometrically, as the material was heated in a sealed vessel.[3] In modern practice however, EGA is almost always conducted simultaneously, or as nearly as possible, with other thermal analysis (TA) techniques, giving rise to e.g. simultaneous TG-EGA, or TG-DTA-EGA. The advantages of simultaneous methods have been described elsewhere,[4] and EGA is particularly valuable in that it provides direct chemical information about the processes followed by the physical techniques of TG, DTA and others.

The driving force for the development of coupled EGA methods arose primarily from this need for chemical interpretation of weight loss processes. Other routes to achieving this, such as analyses of final or inter-mediate products may be subject to uncertainty, as the often highly reactive solids may interact with the atmosphere on removal from the instrument. Certainly, examination of the solid reaction residues may be necessary to complete the description of the processes undergone on heating, but EGA offers a direct, and often simple approach. Most TA experiments are carried out under controlled, flowing atmospheres, and the coupling of some form of gas detector can yield valuable information which would otherwise be thrown away via the gas outlet.

Many other advantages of EGA have emerged, above that of a qualitative description of the reaction. When specific detectors are used, the evolution of one product may be followed against a background of overlapping or concurrent product evolutions, which are seen as a single event by, e.g., TG. Some of the detectors used are extremely sensitive, and may allow lower detection limits than TG or DTA. Some are capable of precise quantitative performance, and this may be especially valuable when complex mixtures of materials are studied, and quantification by, e.g. TG, is not possible. Most of the processes conventionally studied by TG are sensitive to atmosphere quality, and EGA can be invaluable in determining this precisely, or in establishing a suitable purging procedure for a particular instrument.

The only text-book devoted to the subject of EGA is that edited by Lodding,[5] which covers the field up to 1967, and discusses many experimental factors in good detail. Chapters on EGA are to be found in general TA texts.[6,7,8,9] The bulk of the work on EGA remains dispersed in the literature.

The present paper will concentrate on typical current methodology, with emphasis on simultaneous EGA using Fourier transform infrared spectroscopy (FTIR) and mass spectrometry (MS) instruments.

2 INTERFACING CONSIDERATIONS

The successful design of a coupled TA-EGA apparatus has been realised in a multitude of ways. We may distinguish between continuous sampling, in which a signal representing the concentration of a product in the purge gas stream is continuously recorded as a function of time and/or temperature, and intermittent sampling, in which a portion of the evolved gases is collected over a chosen temperature range and then analysed "off-line". In the former category are included wholly analogue devices, as well as those capable of providing an effectively continuous record of the changing nature of the gas stream by generating information at acceptably small time intervals (e.g. FTIR or MS, where a spectrum may be recorded every few seconds). In the latter category gas chromatography (GC), or perhaps GC-FTIR or GC-MS, are obvious examples, in which the frequency of sampling is dictated by the elution time of the products on the chromatograph. More frequent sampling is possible by switching in a series of cold traps, adsorption tubes etc. to the purge gas line, and analysing them later.[10] The use of some form of separation technique, such as GC, may well be required at some stage in the study of materials which evolve a complex range of (particularly organic) products on decomposition.[11]

In all interfaces for EGA, the following points should be considered:

1) The product/purge gas mixture presented to the detector/analyser should be representative of that close to the source of the evolved products.

2) The products should be transmitted to the analyser in as short a time as possible. This is desirable for several reasons: effectively simultaneous recording of the EGA and TA data is necessary to realise the advantages of coupled techniques; secondary gas-phase reactions may alter the product composition, especially where there is a large temperature difference between the points of generation and registration; long time lags, usually because of large volumes in the connecting pipework, can lead to distortions in the EGA profile, and loss of resolution through diffusion broadening.

3) The transfer line should be as inert as possible. Heating may be necessary to avoid adsorption or condensation problems, but could induce degradation, particularly on metal surfaces.

4) The response time of the detector itself should be considered.

5) The nature of the products to be studied, and the range of purge gases to be used, may dictate the form and/or materials of the interface.

6) The optimum flowrates for the TA instrument and the gas analyser may not coincide. It may be necessary to sample only a portion of the evolved gases, or it may be possible to capture them all. In certain cases a "make-up" gas stream may be needed to shorten the response time in larger volume analysers.

7) The performance of the TA instrument should not be compromised, nor should it be necessary to alter the conditions of the experiment to suit the gas analyser.

All the above points interact, and the final design will often be a compromise, but adequate for a particular purpose. Modern commercially available coupled instruments can offer great flexibility and good performance, and are likely to obviate the need for "home-made" apparatuses in many cases.

3   TYPES OF DETECTORS

The value of coupling EGA to TA has long been recognised, and considerable ingenuity has been shown in linking almost every conceivable type of gas detector/analyser to most forms of TA instrument. The following list covers the main types, though is by no means exhaustive.

Indicator paper
Draeger tube
Katharometer [12]
Absorption into solution followed by –
    colourimetry
    pH measurement
    ion specific electrode measurement [13]
    coulometry [14]
    titrimetry [15]
Paramagnetic oxygen analyser
Chemi-luminescent NO_x analyser
Electrolytic detectors for e.g. H_2O, SO_2 [16,17]
Non-dispersive infrared (NDIR) [17]
FTIR [18]
MS [19]
GC (alone, or preceding e.g. FTIR or MS) [20]

The first two on the list, though of limited capability, may be enough to decide between two possible products, and answer the question at hand. Katharometers with high sensitivity were much used in EGD, and in the case of a single product being monitored can perform quantitatively, taking the method into the scope of EGA. Absorption of the products into solution, by passing the gas through a suitable bubbler offers a wide range of possibilities for analysis, mostly potentially quantitative, specific, and capable of providing a continuous record of changing gas composition. Problems may be encountered with the response time of the complete system, or noise transmitted to the balance mechanism due to the bubbling. The specific detectors mentioned for H_2O, SO_2 etc., as well as NDIR cells, are capable of high sensitivity and stability. High gas flowrates may be necessary to preserve short response times, which will offset their sensitivity to some extent.

Most of the detectors mentioned may be linked to the thermobalance or DTA instrument with little difficulty; the use of GC or MS detectors requires more careful thought. The special requirements of the MS will be discussed later. The main consideration in designing a GC sampling arrangement should be to ensure that no pressurisation, and no pressure fluctuations, are allowed to affect the performance of the thermobalance.

In the following sections two specific, single-gas detectors used for EGA will be described, together with examples of their application. The cases of FTIR and MS will then be covered in more detail, as these instruments have recently been strongly favoured.

4   SINGLE-GAS DETECTORS

1) Hygrometer

Many materials evolve water on heating whether by desorption, hydrate dissociation, dehydroxylation or chemical reaction, yet it is notoriously difficult to quantify in small amounts, largely because of its omnipresence. In the course of a study of some hydrated salts, an electrolytic hygrometer was coupled to a low temperature DTA unit[16] for EGA of water.

The coupled units are shown in Figure 1. The water vapour in the gas stream from the DTA is absorbed by a phosphoric acid coating in the hygrometer cell, on fine platinum windings. An electrolysis current continually

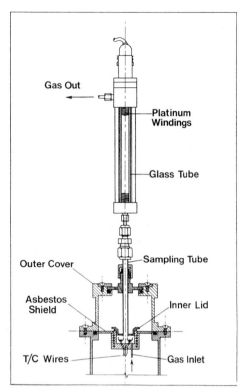

Fig.1 DTA fitted with electrolytic hygrometer. (From ref.16; reproduced by permission of Birkhauser Verlag.)

flows which is proportional to the amount of water in the coating, which in turn is related to the amount of water being absorbed from the gas stream. Good quantitative performance was demonstrated by an excellent correlation between the EGA peak area and the amount of water evolved, in the range 0.05-0.5mg. The evolution of the first four waters of hydration of copper sulphate pentahydrate were used for the calibration

The combination proved invaluable in the interpretation of the DTA curves for barium nitrite monohydrate. This salt undergoes a variety of transitions up to the melting point, the occurrence and nature of which can depend on most applied experimental parameters. One problem concerned the irregular DTA peak at about 175°C, which altered dramatically in form with heating rate, sample preparation etc., or was sometimes absent. DTA and EGA curves are shown in Fig.2 for a sample of fine (<150μm) crystals heated in an open pan. The EGA trace in the region of the DTA peak of interest mirrors it exactly, showing it unambiguously to be associated with water evolution. The difference in form of the traces at about 100°C led to the identification of the sharp DTA peak, not accompanied by water loss, as a phase transition in the partly decomposed hydrate.

2) Non-dispersive infrared analyser

A schematic diagram of a non-dispersive infrared analyser is shown in Fig.3. Purge gas carrying the product CO_2 passes continuously through either of the two analysis cells. Pulses of broadband infrared radiation pass through the analysis and reference cells, when absorption of IR energy by the CO_2 causes a pressure imbalance between the cells. This is detected and converted to an electrical signal proportional to the concentration of CO_2. The instrument has a large

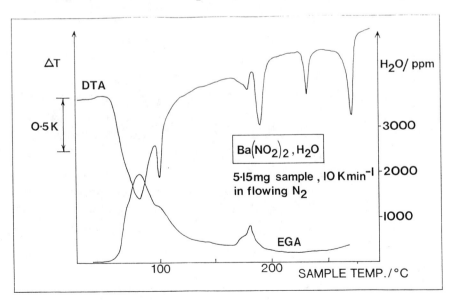

Fig.2 DTA and EGA (water) curves for barium nitrite monohydrate.

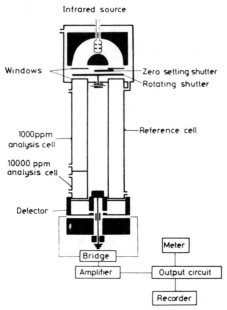

Fig.3   Schematic diagram of NDIR analyser for CO_2 and H_2O. (From ref.17; reprinted by permission of John Wiley and Sons. Ltd.)

volume, but was linked to a DTA unit with a large bore furnace tube to form a highly successful combination.[17] In addition to NDIR for CO_2, a further unit for water analysis and an electrolytic SO_2 analyser were coupled in series. The combination has been, and continues to be, applied to a wide range of mineral studies. The sensitivity was demonstrated by an ability to identify, and to quantitatively determine levels of carbonates in artificial mixtures at levels as low as 50 ppm.[21]

5   FOURIER TRANSFORM INFRARED SPECTROSCOPY (FTIR)
It is interesting to note that in 1967 highly promising results were demonstrated with an FTIR instrument for identifying low concentrations of material in a flowing gas stream.[22] The technique was at that time restricted in application largely through cost and complexity. The field up to 1982-3 has been reviewed by Lephart.[23] Recent technical improvements, together with readily

available increased computing power, have made these instruments fairly common, to the extent that they will largely replace dispersive instruments in the laboratory. The high sensitivity and fast scanning ability of FTIR instruments makes them suitable for EGA.

A coupled TG-FTIR system was developed at Standard Oil by Mittleman et al.[24] which formed the basis of the first commercially available combined system.[18] Evolved gases from a Stanton Redcroft TG1500 thermobalance, which has a furnace volume of about $12cm^3$, were swept from the sample under slight pressure, at a flowrate of $20-25cm^3$/min., via a 0.7mm I.D. glass-lined stainless steel tube, into the analysis cell shown in Fig.4. The cell has a low volume, comparable with that of the furnace, and flowrate matching with good resolution was accomplished. The path of the IR beam is shown, passing through the "double-glazed" arrangement with corrosion-resistant inner ZnSe windows, which assists in minimising condensation of volatiles. The entire cell and transfer line may be heated.

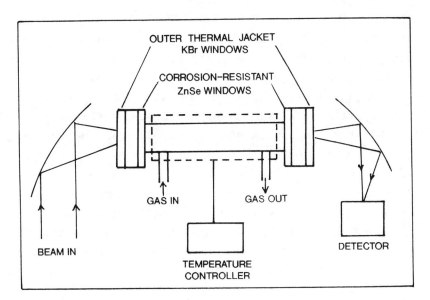

Fig.4 Schematic diagram of a gas cell for EGA by FTIR.

Some results from a study on calcium oxalate decomposition are given in figures 5 and 6. Fig.5 shows spectra taken during the three distinct weight loss stages, which identify the products as water, $CO+CO_2$ and CO_2. Fig.6 shows how the computer can be made to reconstruct EGA profiles for these gases from the stored scans (here taken at 15s intervals) by nominating an appropriate spectral region in which they absorb radiation. Fig.6d is termed a Gram-Schmidt reconstruction, and measures the integrated IR absorption during the experiment. It thus may be compared to the DTG curve in Fig.6e. Large differences in sensitivity are seen for the three gases.

Another approach to interfacing TG and FTIR instruments is illustrated by the Du Pont (TA Instruments)/Nicolet combination. Here, the ground glass ball-joint on the furnace tube of the model 951 thermobalance makes coupling to the specially-designed cell particularly simple.[25]

6   MASS SPECTROMETRY (MS)

The use of mass spectrometers for EGA has a rich and comparatively long history. For an introduction to the subject the reader is referred to the excellent reviews by Langer and Gohlke[26], Friedman[27] and Holdiness.[28]

A proper description of the means of operation of mass spectrometers is beyond the scope of this article, but all have in common some method of ionising incoming gas molecules, usually under high vacuum, the filtering of the ionised species by an analyser (magnetic sector, time-of-flight or quadrupole) and the presentation of these species to a detector, so that a mass spectrum is obtained - i.e. a record of the amount of the species at a given mass number. The advantage of MS is the ability to detect all gaseous species, usually at very high sensitivity.

The essential problem in using MS for EGA work lies in the large pressure difference between the TA unit,

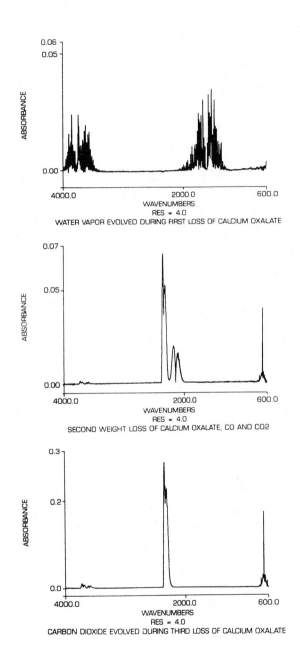

Fig.5 FTIR spectra recorded during the three stages of decomposition of calcium oxalate monohdrate. (5mg sample, 20°Cmin.$^{-1}$, nitrogen atmosphere.

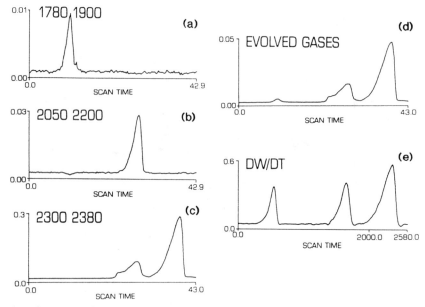

Fig.6   Evolved gas profiles for H_2O, CO and CO_2 reconstructed from stored spectra, together with the Gram-Schmidt plot and the DTG curve.

nominally at atmospheric pressure, and the high vacuum in the ion source (typically a factor of 10^9). Mention should be made however of certain chemical ionisation instruments, in which the source pressure can be as high as atmospheric, and interfacing is more direct.[29]

The course of development of EGA-MS is charted by the following sequence of methods used:

1)   Heating the sample under vacuum in or near the ion source[30]

2)   Building TG[31] or DTA[32] sensors into the high vacuum enclosure of the mass spectrometer

3)   Linking a thermobalance capable of vacuum operation to the mass spectrometer[33]

4)   Linking a conventional thermobalance to the mass spctrometer[19]

In the context of "mainstream" TA, it is the last method which is of most interest, though the others will be required in some circumstances.

There are many devices for effecting the drop from atmospheric pressure to ca.10^{-6}mbar. The chief ones (see e.g.McFadden,[34] Gudzinowicz & Martin[35]) are:(1)Leak valve;[36] (2)Orifice(s);[37] (3)Effusion separator;[34] (4)Membrane separator;[38] (5)Jet separator;[39] (6)Capillary with bypass.[38]

In each case only a small portion of the purge gas stream enters the ion source, so that the high vacuum may be preserved. The bulk of the gas is pumped away. The devices termed "separators" in the list perform some enrichment of the products in the stream to be analysed, by separating out some of the purge gas. The enrichment may be performed by virtue of a density difference (jet,effusion), or by a difference in the rate of diffusion through a membrane. The other methods in ideal circumstances deliver a fixed, and representative fraction of the gas mixture to the ion source. Requirements for the satisfactory accomplishment of this were discussed by Korobeinichev[40] and also by Dollimore et al.,[41] where their chief concern was the possible distortion of the EGA curves when used for kinetic studies.

A dual inlet system used by the author for a number of years is shown in Fig.7.[42]

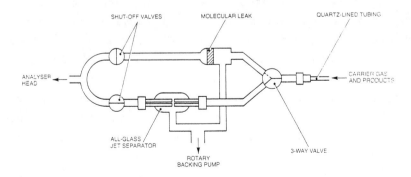

Fig.7 Dual inlet system for mass spectrometer.

The jet separator operates with a low density purge gas (typically helium, though hydrogen is possible) and allows most of the products to pass to the ion source, while the helium diffuses from the fine stream between the two jets and is pumped away. Heavier molecules are transmitted more efficiently than light ones, and the device thus shows mass discrimination, which may be allowed for by suitable calibration procedures. The sensitivity of this interface is some two orders of magnitude higher than the other, the capillary with bypass, but this can operate with any purge gas and is thus far more flexible. The sensitivity of the latter is still adequate for the majority of applications (ca 90% of the varied work in the author's laboratory). A drawback of the system in Fig.7, which incorporates metal valves and pipework, is the inefficient transmission of strongly acidic gases (e.g. HCl), which has been overcome by a continuously-pumped inlet,[43] with only quartz between the sample and analyser.

All these inlets draw in purge gas/product mixture at a fixed rate (depending on the nature of the gas, and the temperature of the capillary), which is typically of the order of $10cm^3$/min. This is at the low end of the range of flowrates suitable for most TA units, and an initial split is required. The arrangement used for the Stanton Redcroft STA1500 TG-DTA unit is shown in Fig.8.

A flexible, heated, silica-lined stainless steel capillary is connected via a short alumina tube to the chamber surrounding the sample. The small swept volume (ca.$4cm^3$) used with a typical gas flowrate of $50cm^3$/min gives good resolution of sharp EGA peaks, and efficient purging of the area around the sample. The excess purge gas, not drawn by the capillary, passes through the normal exit route. In this way pressurisation or evacuation of the instrument is avoided. The resolution of the combined system, and an indication of the lag between product generation and registration, have been illustrated by the near coincidence both in time and

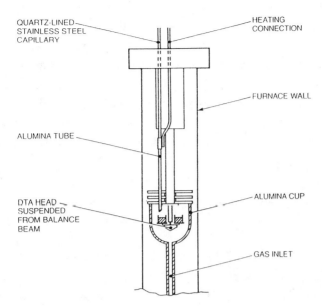

QUARTZ-LINED STAINLESS STEEL CAPILLARY

HEATING CONNECTION

FURNACE WALL

ALUMINA TUBE

DTA HEAD SUSPENDED FROM BALANCE BEAM

ALUMINA CUP

GAS INLET

Fig.8 Gas sampling arrangement for Stanton Redcroft STA1500 TG-DTA unit.

shape, of small DTA and EGA peaks arising from the bubbling of a fused sample.[44] The delay between the instruments is estimated as < 0.5s. Similar arrangements are used for other instruments.[39,43] though a ceramic orifice system for high-temperature work with products of low volatility has been described.[37]

As for the analyser itself, the quadrupole type is now used almost exclusively. Its robustness, relatively low cost, and ease of use compared to a magnetic sector instrument together justify its acceptance for dedicated EGA use by non-specialists in mass spectrometry. Most manufacturers now offer MS couplings for at least some of their equipment, though the control and analysis software for the TA and MS units is not always satisfactorily integrated.

The mode of operation of the author's system, which uses a VG Gas Analysis 300amu range quadrupole MS, may be illustrated by the decomposition of calcium oxalate, when the parallels with the FTIR approach may be seen. The instrument may be set to record complete mass scans

every few seconds on a fixed sensitivity, or, somewhat
more slowly, with automatic gain ranging for each mass
number. This mode is of value in screening runs on an
unknown material. Scans collected during the three main
weight loss regions are shown in Fig.9. The presence of
water, CO and CO_2 are revealed by the peaks
corresponding to the molecular masses of these species,
while the other, major peaks arise from the argon purge
gas. EGA curves for chosen mass numbers may be
constructed from the scans, or for more accurate
monitoring of the EGA peaks, the instrument can be
arranged to record only the signals characteristic of
these species with a typical resolution of ca. 1s
(Fig.10).

   Two applications illustrate various aspects of the
TA-MS technique. The first is from a study of coal
combustion.[45] A typical experiment on a bituminous coal
is represented in Fig.11. The sample was examined in
air, while monitoring continuously a range of gases
including those shown which are water(18amu), NO(30amu),
CO_2(44amu) and SO_2(64amu). The exotherm at 330-350°C
corresponds to the oxidation of released volatile
matter, and precedes the larger exotherm due to char
combustion at 490-520°C.

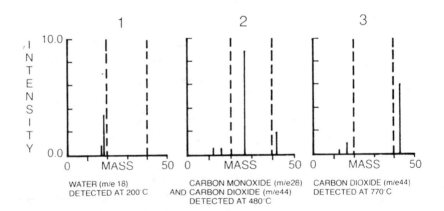

Fig.9 Mass spectra recorded during the three stages of
decomposition of calcium oxalate monohydrate.

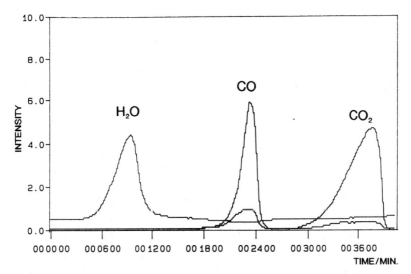

Fig.10   EGA curves for H_2O, CO and CO_2 obtained by selected ion monitoring.

It is evident from the H_2O profile that the majority of the hydrogen-containing material is oxidised in the first stage. It is interesting to note that a substantial proportion of the heat output, and a considerable amount of gas, is produced during a region of weight increase, caused by chemisorption of oxygen. Evolution of NO reached a maximum in the region of 560°C which shows the high stability of the organic nitrogen structures in the coal. The SO_2 profile shows peaks due to pyrite oxidation and combustion of organic sulphur.

The high sensitivity of the jet separator inlet was used in a study of a polyimide resin blend.[44] In Fig.12 the DTA curve shows an irregular melting peak at about 110°C, after at least two glass transitions. An apparently simple, symmetrical curing exotherm follows, with a weight loss of some 7%. The EGA traces (43amu, acetone; 105amu, acetophenone; 26amu, acetylene and 114amu, unassigned) show that some reaction occurs during the melt. The weight loss in this region is less than 0.2%, yet the EGA traces are able to provide detailed information on processes which might have been

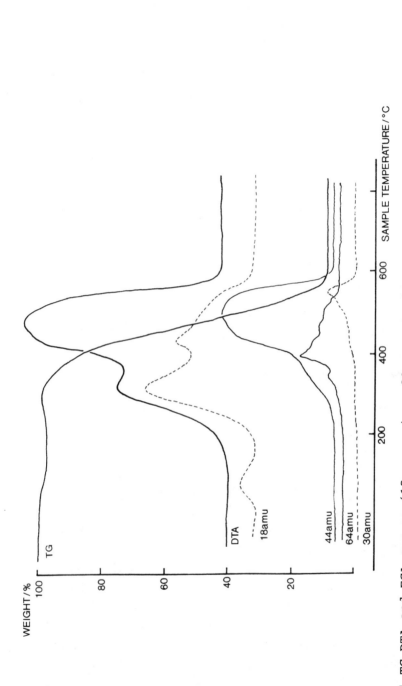

Fig.11 TG,DTA and EGA curves (18amu, water; 30amu, NO; 44amu, CO_2; 64amu, SO_2) for a bituminous coal. (15mg sample, 15°Cmin.$^{-1}$, flowing air atmosphere)

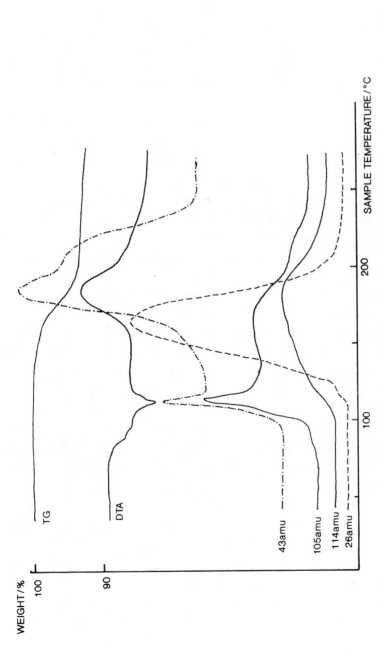

Fig.12 TG, DTA and EGA curves (26amu, acetylene; 43amu, acetone; 105amu, acetophenone and 114amu, unassigned) for an uncured polyimide resin. (15mg sample, 20°Cmin.$^{-1}$, helium atmosphere)

overlooked. The curing reaction is far from simple; only four gases are shown in the figure, but four others recorded during the run each showed some individual feature.

7 DISCUSSION

In recent years attention has focused upon FTIR and MS instruments for EGA work. This has been due to a number of factors:

1) technical developments – both in FTIR and quadrupole MS – have led to the availability of commercial devices at accessible prices

2) both are usually computer-controlled and require little operator intervention during experiments

3) both are capable of recording a complete spectrum in a short time (seconds)

4) both can be extremely sensitive, though their relative sensitivity can vary enormously with respect to different products and/or purge gas natures

5) computerised handling of the recorded data has greatly speeded interpretation; spectral libraries may be available for identification of products.

A choice between the two could only be made with regard to the type of work being undertaken. MS is considered to be the more versatile, with the ability to detect non-polar m‍ olecules (O_2, H_2, N_2 etc.) which FTIR can not. The inherent sensitivity of detection is high for MS, and does not vary greatly (ca. factor of 10) for most gaseous products. The usable sensitivity however is limited, particularly for water, by the instrument background signal. Most MS interfaces involve a capillary tube at some stage, and this could be susceptible to blockage by certain classes of products, such as smokes, tars or materials which re-polymerise. A surprisingly wide range of samples has however been successfully examined.[43] The interfaces for FTIR are of a more open nature, and are more suited to handling difficult products, though frequent cleaning may be necessary. The relative sensitivity of FTIR can vary by a factor of 1000 or more for different volatiles, though the intrinsic sensitivity of the FTIR technique may be able to cope with this, giving reasonable detection limits for weakly-absorbing species.

Both MS and FTIR should be capable of quantitative operation as EGA instruments, but most work has been of a qualitative or semi-quantitative nature. Though this has been of great value, the establishment of improved experimental procedures and/or calibration techniques should allow precise quantification of products at low concentrations. Much work remains to be done in this area.

8 CONCLUSIONS

The technique of EGA has a diverse range of applications which has only been hinted at in the present article. From its origins in EGD, EGA has progressed to provide specific chemical information, sometimes with high sensitivity and sometimes quantitatively. The two currently most favoured techniques, MS and FTIR, have recently rekindled interest in the subject. The next few years, it is believed, will see the establishment of better quantitative methods for EGA using MS and FTIR, techniques which already have sufficient sensitivity, specificity and versatility to answer most of the questions asked of them by the thermal analyst. It may be at that stage that EGA can be said to have come of age.

9 REFERENCES

1) R.C.MacKenzie in 'Treatise on Analytical Chemistry', Ed. P.J.Elving, 2nd Edition, Part 1, Vol 12, John Wiley, 1983, 1.
2) W.Lodding and L.Hammell, Rev.Sci.Instr., 1959, 30, 885.
3) W.E.Garner (ed.), 'Chemistry of the Solid State', Butterworths, London, 1955.
4) E.L.Charsley, this publication.

5)   W.Lodding (ed.), 'Gas Effluent Analysis', Marcel
     Dekker, New York, 1967.
6)   M.E.Brown, 'Introduction to Thermal Analysis
     Techniques and Applications', Chapman and Hall,
     London, 1988.
7)   P.D.Garn, 'Thermoanalytical Methods of
     Investigation', Academic Press, New York, 1965.
8)   W.W.Wendlandt, 'Thermal Analysis (3rd.edn.)', Wiley,
     New York, 1986.
9)   F.Paulik and J.Paulik in: C.L.Wilson and D.W.Wilson
     (eds.), 'Comprehensive Analytical Chemistry',
     vol.XII, part A, Elsevier, Amsterdam, 1981.
10) T.-L.Chang and T.E.Mead, Anal.Chem., 1971, 43(4),
     534.
11) P.A.Barnes, G.Stephenson and S.B.Warrington,
     J.Therm.Anal., 1982, 25, 299.
12) W.Lodding and L.Hammell, Anal.Chem., 1960, 32, 657.
13) T.R.Fennell, G.J.Knight and W.W.Wright, in:
     H.G.Wiedemann (ed.),'Thermal Analysis',vol.1,
     Birkhauser, Basel, 1971, 245.
14) S.Brinkworth, R.J.Howes and S.E.Mealor, in:
     D.Dollimore (ed.),'Proceedings 2nd ESTA Conference',
     Heyden, London, 1981.
15) J.Paulik and F.Paulik, Thermochim.Acta, 1972, 4,
     189.
16) S.B.Warrington and P.A.Barnes, in: D.Dollimore
     (ed.), 'Proceedings 2nd.ESTA Confce.',Heyden,
     London, 1981.
17) D.J.Morgan, J.Therm.Anal., 1977, 12, 245.
18) D.A.C.Compton, International Labmate, 1987, 12, 37.
19) F.Zitomer, Anal.Chem., 1968, 40, 1091.
20) H.G.Wiedemann in: R.F.Schwenker and P.D.Garn (eds.),
     'Thermal Analysis',vol.1, Academic Press, New York,
     1969.
21) A.E.Milodowski and D.J.Morgan, Nature, 1980, 286,
     248.
22) M.J.D.Low in: W.Lodding (ed.),'Gas Effluent
     Analysis', Marcel Dekker, New York, 1967.
23) J.O.Lephart, Appl.Spectroscopy Rev., 1982-3, 18,
     265.
24) M.L.Mittleman, D.A.C.Compton and P.Engler, in:
     'Proceedings 13th NATAS Conference', Philadelphia,
     PA, 1984, 410.
25) R.C.Wiebolt, American Laboratory, Jan.1988, 70.
26) H.G.Langer and R.S.Gohlke in W.Lodding (ed.),'Gas
     Effluent Analysis',Marcel Dekker, New York, 1967.
27) H.L.Friedman, Thermochim.Acta, 1970, 1, 199.
28) M.R.Holdiness, Thermochim.Acta, 1984, 75, 361.
29) S.M.Dyszel, Thermochim.Acta, 1983, 61, 169.
30) P.D.Zemany, Anal.Chem., 1952, 24, 1709.
31) D.E.Wilson and F.M.Hamaker in: R.F.Schwenker and
     P.D.Garn (eds.), 'Thermal Analysis', vol.1, Academic
     Press, New York, 1969, 517.
32) H.G.Langer and T.P.Brady, Thermochim.Acta, 1973, 5,
     391.
33) H.G.Wiedemann, Chem.Ing.Tech., 1964, 36, 1105.

34) W.H.McFadden, 'Techniques of combined GC-MS', Wiley, 1973.
35) B.J.Gudzinowicz et al., 'Fundamentals of combined GC-MS', part III, Marcel Dekker, 1977.
36) J.Chiu and A.J.Beattie, Thermochim.Acta, 1980, 4, 251.
37) W.D.Emmerich and E.Kaisersberger, J.Therm.Anal., 1979, 17, 197.
38) M.L.Aspinal, H.J.Madoc-Jones and E.L.Charsley in: H.G.Wiedemann (ed.), 'Thermal Analysis', vol.3, Birkhauser, Basel, 1971, 303.
39) P.A.Barnes, G.Stephenson and S.B.Warrington in: D.Dollimore (ed.), 'Proceedings 2nd ESTA Conference', Heyden, London, 1981, 47.
40) O.P.Korobeinichev, Russ.Chem.Rev., Dec.1987, 957.
41) D.Dollimore, G.A.Gamlen and T.J.Taylor, Thermochim.Acta, 1984, 75, 59.
42) E.L.Charsley and S.B.Warrington, Thermochim.Acta, 1987, 114, 47.
43) E.L.Charsley, S.B.Warrington, G.K.Jones and A.R.McGhie, American Laboratory, Jan.1990.
44) E.L.Charsley, M.R.Newman and S.B.Warrington in: 'Proceedings 16th NATAS Conference', Washington, 1987, 357.
45) P.Burchill, D.G.Richards and S.B.Warrington, Fuel, 1990, 69, 950.

Thermomechanical Analysis and Dynamic Mechanical Analysis

M. Reading

ICI PAINTS, RESEARCH DEPARTMENT, WEXHAM ROAD, SLOUGH, SL2 5DS, BERKS, UK

1 INTRODUCTION

Thermomechanical methods, particularly Dynamic Mechanical Analysis, have grown in sophistication and acceptance to the point where there is virtually no aspect of polymer and composite materials science where they do not play an important role. Their range of applications extends from measuring simple basic quantities such as coefficients of expansion and glass transition temperatures (Tg's) to helping define the morphology of complex, phase separated polymer systems and characterising cross-linking behaviour. Here we attempt to illustrate the many different possible measurements that can be made with these techniques using, where possible, examples from the recent literature.

In practice the vast majority of these instruments are used to study polymers or polymer containing systems. This article reflects this bias but it should be borne in mind that any material that exhibits viscoelastic behaviour, crystallinity, glass transitions etc. can be studied using these techniques.

2 BASIC PRINCIPLES

All thermomechanical instruments consist of a means of applying stress in a controlled way while measuring strain (or vice versa) together with a means of controlling sample temperature both isothermally and, almost always, as a temperature ramp. Thermomechanical analysers (TMA's) traditionally apply a static stress and measure strain as a function of either time or temperature. However, modern instruments, illustrated schematically in Fig.1, can carry out controlled strain experiments and even stress ramps. When the sample is not deformed by the applied stress the strain simply becomes a gauge of sample size along the measured axis as a function of, usually, temperature. This is the basis of dilatometry.

Dynamic Mechanical Analysis (DMA) requires that the stress or strain being applied is varied in a periodic way as a function of time, usually sinusoidaly. The design of commercially available instruments is diverse and no single schematic diagram could represent them all. Fig.1 also represents the design of one of the most recently introduced commercial DMA's, but many of the others are significantly different. However, in essence, they are all similar insofar as they provide a method of applying a combination static and dynamic stress and measuring strain (or vice versa). Most often it is the dynamic strain that is held constant. For thin films or fibres held in tension there is a requirement that an additional constant stress be applied so that there is a residual force keeping the sample under tension at all points during the deformation cycle. This is to avoid buckling the sample. This additional stress needs to be changed in concert with the dynamic stress as the modulus of the sample changes, this usually produces a near constant overall strain.

It should be noted that there are many possible types of deformation, including shear, flexure, tension and torsion.

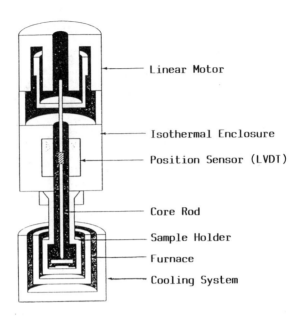

Figure 1   Schematic Diagram of a TMA/DMA (Reprinted with
kind permission of Perkin Elmer)

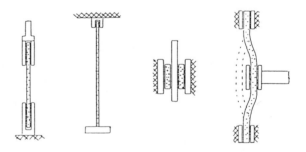

Figure 2   Modes of Deformation. Samples Given Spotted Shading.
Left to Right; Tension, Torsion, Shear and Flexure.

These are illustrated in Fig.2. The choice of an appropriate
type of deformation depends upon the type of material being
considered and the type of information required. For example,
tension is clearly more appropriate for a fibre or a thin film
than flexure. Flexure is often preferred for stiff composites
as this mode usually requires less force to obtain a
measurable deformation. This is important because, when large
forces are needed to obtain a measurement, instrument errors
due to yielding of the components of the measuring system
itself can become significant. For anisotropic materials
mounting the sample in different ways and using different
modes provides information on the nature of the anisotropy.
From the point of view of measuring the temperatures of
transitions and the mechanical properties of isotropic
materials, all of the different types of deformation should
provide essentially the same type of information.

Polymeric materials exhibit markedly viscoelastic
behaviour which simply means that they exhibit a combination
of the ability to recover elastically when deformed and the
ability to flow. This is generally expressed in terms of a
dynamic storage modulus, denoted G' for shear deformation,
which quantifies the material's ability to store energy
(elastic behaviour) and a loss modulus, denoted G'', which
measures the material's ability to dissipate energy by flow
(viscous behaviour). In Dynamic Mechanical Analysis where the
periodic deformation can be expressed as a sine wave, G' is
measured by quantifying that part of the sample's resistance
to deformation that is in phase with the applied stress, while
G'' is determined from the component of the resistance that is
out of phase. This is usually represented as an Argand diagram
as illustrated in Fig.3 which shows how the response of the
sample is split into real (in phase) and imaginary (out of
phase) components thus, for shear measurements;

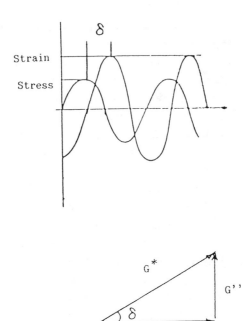

Figure 3   Analysis of Dynamic Mechanical Results

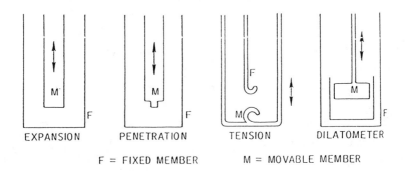

EXPANSION   PENETRATION   TENSION   DILATOMETER

F = FIXED MEMBER   M = MOVABLE MEMBER

Figure 4   TMA Probe Configurations (Reprinted by Courtesy of
TA Instruments)

$$G^* = G' + G''i$$

Where $\quad G^* = $ the dynamic shear modulus

$\qquad\quad G' = $ the storage modulus

$\qquad\quad G'' = $ the loss modulus

and $\qquad \tan\delta = G''/G'$

Where $\quad \delta = $ the phase angle lag

The properties most usually plotted from a DMA experiment are storage modulus and tanδ as a function of temperature.

When a constant stress is applied, as in TMA, the viscoelasticity manifests itself as an instantaneous (elastic and recoverable) deformation followed by a creeping (viscous and unrecoverable) deformation. This type of experiment is termed creep recovery or creep relaxation. The analog when a constant strain is applied is called stress relaxation. The relaxation behaviour of polymers and how it changes as a function of time and temperature, ie. the relaxation kinetics, can be described by an equation known as the WLF (William Landel Ferry) equation. Application of this equation can make it possible to predict the long term behaviour of polymers by carrying out a comparatively rapid series of laboratory experiments[1].

While TMA's and DMA's currently form two fairly distinct families of instruments, modern equipment is blurring the distinction by effectively combining some or all of the capabilities of both types of instrument. A feature of modern TMA and DMA instruments is the large number of different attachments that can be bought to enable a single instrument to operate in a variety of different modes. While instruments producing a torsional deformation are usually dedicated to this mode exclusively, flexure, tension and shear can often be provided by one instrument simply by changing the clamping system. For TMA's there are also special probes such as penetration probes that have small area rounded ends. A piston

like cell can be used to make dilatometric measurements on
finely divided samples as well as the more usual uniaxial
expansion measurements made with a flat probe. Large area flat
probes can be used to form a parallel plate rheometer. Often
users can design and fit their own custom made probes for
special applications. Fig.4 gives the most commonly used
types.

3 APPLICATIONS

An attempt is made to summarise the applications of the
various modes of TMA and DMA in a tabular form in tables 1 and
2. These are explored in more detail below.

TMA
The flat probe for the TMA can be used to measure not only
coefficients of expansion but also glass transition
temperatures as the rate of expansion increases sharply as a
polymer goes from a glass to a rubber or fluid. The same
comment applies to the piston-like dilatometer cell used for
making measurements on powder samples. Fig.5 shows a
measurement made on powdered polystyrene with this device.
    The pointed probe can be used to measure softening
temperatures. Fig.6 shows a typical result where the probe
penetrates into the sample as it passes through its glass
transition. The softening temperature is taken to be the
extrapolated onset of this process. Some caution must be
exercised to distinguish between the measurement of a Tg and
the measurement of a softening temperature. Though the two are
clearly related, the latter is very dependent on the applied
load and the shape of the probe point etc.. However, provided
the experimental conditions are maintained constant, this
measurement can be used to rank Tg's for a series of samples.
Fig.6 shows how it can be used to investigate degree of cure[2].
The use of the TMA in this way can also provide a very

Method & Sample	Mode	Quantity Measured	As a Function of	Information Obtained
TMA bulk sample	flat probe with light load	compression	Temperature	Coefficient of expansion and Tg
TMA divided sample	Dilatometer attachment	Volumetric expansion or shrinkage	Temperature	Coefficient of expansion and Tg
TMA thin film	Penetration probe with significant load	Depth of penetration	Force	Modulus cross-link density
			Time	Creep behaviour Cure behaviour
			Temperature	Softening point (Tg) Melting point
TMA film or fibre	Film/fibre tension attachment	Uniaxial extension or shrinkage	Force	Modulus cross-link density
			Time	Creep behaviour
				Cure behaviour
			Temperature	Tg, Melting point, Cure behaviour Preparative history
TMA fluid	Parallel plate probe	Distance between plates	Time	Viscosity Gelation
			Temperature	Melting point Viscosity Gelation
TMA Bulk or supported sample	Flexure probe	Degree of bending	Time	Creep behaviour
			Temperature	Softening point (Tg), Melting point

Table 1 Summary of TMA Applications

Method & Sample	Mode	Quantity Measured	As a Function of	Information Obtained
DMA Bulk or Films	Flexure Tension Torsion Compression Shear With oscillatory stress/strain	Force as function of displacement and phase angle lag	Time	Dynamic modulus Cross-link density Cure Behaviour
			Temperature	Tg and other transitions Morphology Melting point Crystallisation Cure behaviour
			Frequency & Temperature	Relaxation behaviour (kinetics)
DMA Supported	Torsion Shear Flexure With oscillatory stress/strain	As for non Supported	As for non Supported	As for non Supported +
			Time & Temperature	Full cure character-isation Gelation Vitrification Cure kinetics
DMA Bulk Films Supported	Tension Shear Flexure With stepwise stress/strain	Stress or strain as a function of time	Time	Creep or compliance behaviour
			Time & Temperature	Full relaxation behaviour WLF equation for predicting long term behaviour

Table 2 Summary of DMA Applications

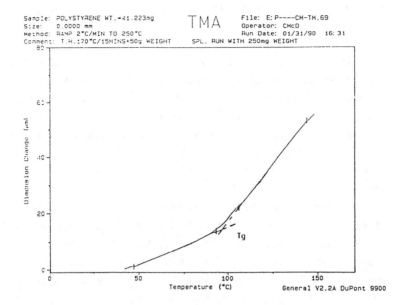

Sample: POLYSTYRENE WT.=41.223mg
Size: 0.0000 mm
Method: RAMP 2°C/MIN TO 250°C
Comment: T.H.170°C/15MINS+50g WEIGHT

TMA

File: E:P----CM-TM.69
Operator: CMcD
Run Date: 01/31/90 16:31
SPL. RUN WITH 250mg WEIGHT

Figure 5   TMA Results For Polystyrene Obtained Using a
Dilatometer Probe

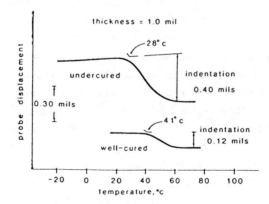

Figure 6   TMA Results For an Acrylic Coating in Penetrometer
Probe (From 2)

convenient means of, for example, studying the efficiency of plasticisers[3]

A further measurement that can be made with a penetrometer probe is that of hardness. By measuring the depth of penetration for a given load a comparative measure of harness can be made. If the Poisson's ratio for the material is known the Hertz equation[4] can be used to determine the modulus of a sample. Measurements of modulus made in this way do not always agree very well with values obtained by other techniques, they do provide a quick and easy way of ranking systems[2].

Constant stress tensile measurements made using TMA are extensively used for studying fibres. The way in which a fibre contracts or expands as a function of temperature can provide a great deal of information on the thermal history of the sample, the frozen-in stress, the degree of crystallinity, the glass transition temperature and the sample's melting point[5]. Another approach that is thought to provide some advantages is constant strain TMA or Thermal Stress Analysis. Recent work[6] has shown how a model postulating two types of domain in yarn, a crystalline domain and an amorphous inter-crytallite domain, can be fitted very accurately to Thermal Stress Analysis data. In this way details of the morphology of the yarn are revealed and the effect of processing conditions can be investigated.

Fig.7 shows a stress scan performed on a paint film. It provides a value for the modulus, the yield point and the strength of the material. Similar experiments can be used to investigate fracture behaviour. These types of measurements have traditionally been made using specialist tensile testers. The fact that they can now be made with TMA is a good illustration of how the increasing versatility of these instruments is expanding their range of applications.

Large area flat probes have been used to study the rheology of thermoset cure[7] but this type of investigation is now more usually undertaken using torsional rheometers with hot stage attachments or supported DMA techniques (see below).

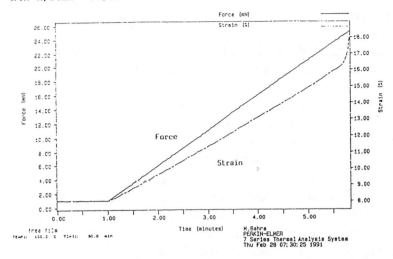

Curve 1: DMA Creep Recovery in Extension
File Info: d-on-1a :5Wed Feb 27 18:32:33 1991
Sample Height: 9.853 mm Creep Stress: 500.0 mN

H.Bahra
PERKIN-ELMER
7 Series Thermal Analysis System
Thu Feb 28 07:30:25 1991

<u>Figure 7</u> TMA Results For a Thin Film Subjected to a Stress
Scan

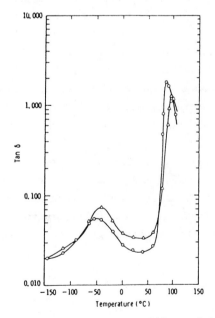

<u>Figure 8</u> DMA Result Showing the Effect of Adding Different
Levels of Rubber to a Cross-Linked Matrix (From 8)

DMA

As a polymer passes through its Tg its tanδ goes through a
maximum and the storage modulus usually decreases by about two
to three orders of magnitude. DMA, therefore, provides the
most sensitive means of studying these types of transition of
all thermoanalytical methods. The inclusion of low Tg domains
within a brittle higher Tg matrix is a well established method
of obtaining tough materials. DMA is often used to examine the
degree of phase separation when two polymers are mixed
together in this manner. Where two distinct Tg's are found
that have the same values as the pure materials then complete
phase separation can be inferred. If the peaks become less
well differentiated and move toward one another then there
must be some degree of miscibility. With care more information
about the way the components in a multi-component system
interact can be gained. For example, there is evidence that
the magnitude of the lower temperature tanδ peak can be
correlated with the fracture energy for some toughened
systems. Fig.8 illustrates how the amount of rubber component
influences the height of the lower temperature transition for
a cross-linked epoxy resin modified with a carboxyl-terminated
acrylonitrile-butadiene rubber, this has been correlated with
fracture properties[8]. Also the strength of adhesion between
the glass filler and the polymer matrix in a fibre-glass
reinforced composite can be correlated with the magnitude of
tanδ at the glass transition[9].

An important feature of cross-linking systems is that of
gelation, the point at which a true network is first formed
and the viscosity rises to infinity. This is then followed by
a build up in cross-link density and consequently storage
modulus. A simple equation for describing this behaviour is[10];

$$G = dRT/Mc$$

Where

G = the equilibrium storage modulus

d = the density

R = the gas constant

T = the absolute temperature

Mc= the average molecular weight between
cross-links

In this way the mechanical properties measured by DMA (the dynamic storage modulus will approximate to the equilibrium storage modulus at low frequencies) and TMA can be related to polymer structure.

DMA is frequently used to monitor the 'cure' of such systems. Because they start as a liquid and are consequently not suitable for use in a DMA, a support is used. Most often this support is a glass fibre braid or ribbon that is impregnated with the sample at the start of the experiment. The subsequent gelation and build up in cross-link density causes the polymer-braid composite to stiffen and provide measurable results. Fig.9 shows some recent results from a study of the curing of an isocyanate-hydroxy system at various heating rates co-plotted with the results of a mathematical model of the curing process that incorporates the equation given above[11]. It can be seen that the DMA results clearly show the gel point and subsequent build up in modulus and, therefore, cross-link density in broad agreement with the model. From these results kinetic parameters like the activation energy can be calculated[11].

Another supported technique consists of using thin metal shim. A coating is spread on the surface of the shim which is then placed in the DMA in a three point bend configuration. Fig.10 shows the results of an experiment carried out on an emulsion paint. The drying behaviour, solvent loss followed by close packing and coalescence of the latex particles, can be clearly followed on the storage modulus curve. The probe

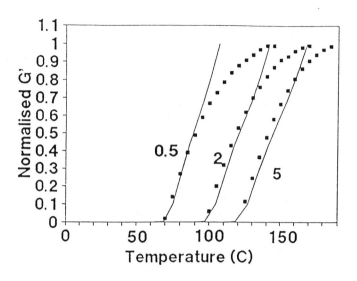

<u>Figure 9</u>   DMA Result Showing the Cure of a Cross-Linking
System at Different Heating Rates (Squares)
Plotted With the Results From a Computer Model
(Lines)

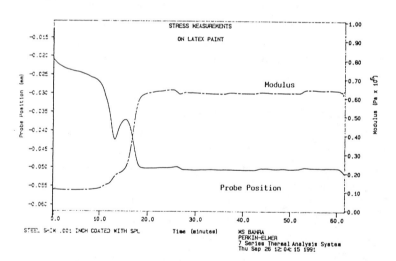

<u>Figure 10</u> DMA Result Showing the Drying Behaviour of a
Latex Paint Including the Build-Up of Internal
Stress (From 13)

position indicates the degree of curvature of the coated shim and therefore serves as a measure of internal stress building up within the coating. This information is useful for studying cracking behaviour in paint systems[12].

Engineering polymers are often semi-crystalline materials. While DSC is the most frequently used method for studying crystallisation processes in polymers, it is equally important to know how these affect mechanical properties. Fig.11 shows the results of DMA scans on a number of commercially available products. It provides at a glance a comparison of their mechanical properties. The effects of processing, nucleating agents etc. can be determined by parallel DSC and DMA experiments[13] in which process conditions are simulated in the laboratory.

4 FUTURE TRENDS

The trend toward more versatile instruments that enable all of the types of measurements outlined above to be carried out in a single instrument will continue and the distinction between TMA and DMA will probably become less important. Combined techniques such as DMA with dielectric analysis and optical measurements are already emerging[14,15] and will grow in importance. Other possibilities include combined DMA and FTIR or FT-Raman, also DMA-Evolved Gas Analysis. The application of TMA and DMA to new ceramic materials will increase, high power DMA's are already appearing to cater for this market. This author hopes that there will be growing interest in the application of Controlled or Constrained Rate Thermal Analysis principles to DMA instruments[16].

5 CONCLUSIONS

Even from the brief and incomplete descriptions given above, it can be seen that thermomechanical methods provide information on coefficients of expansion, modulus (hardness),

damping, glass transitions, morphology, degree of
cross-linking, crystallinity and coalescence. An article such
as this can only scratch the surface of what is an ever
growing field that will become increasingly important in
materials science.

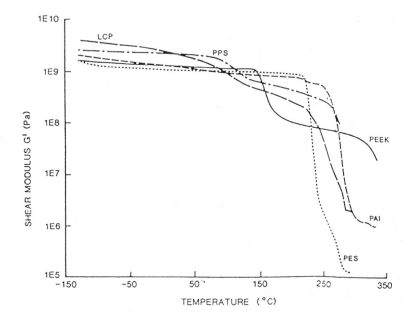

Figure 11 DMA Results For a Variety of Engineering Polymers
          (From 13)

Acknowledgements

The author would like to thank David Elliott for his helpful
comments in the preparation of this manuscript and those
people who kindly gave their permission for material to be
reprinted.

REFERENCES

1. J.J. Aklonis and W.J. MacKnight, 'Introduction to Polymer Viscoelasticity', Wiley and Sons, New York, 1983

2. D.J. Skrovanek and C.K. Schoff, Progress in Organic Coatings, 1988, 16, 135

3. F.C. Masilungan and N.G. Lordi, International Journal of Pharmaceutics, 1984, 20, 295

4. S. Timoshenko and J.N. Goodier, Theory of Elasticity, McGraw-Hill, New York, 1951

5. E.A. Turi Editor, 'Thermal Charaterisation of Polymeric Materials', Academic Press, New York, 1981

6. L.A. Dennis and D.R. Buchanan, Textile Research Journal, 1987, 625

7. L.C. Cessna, H. Jabloner, Journal of Elastomers and Plastics, 1974, 6, 103

8. D.J. Lin, J.M. Ottino and E.L. Thomas, Polymer Engineering and Science, 1985, 25, 1155.

9. P.S. Chua, Polymer Composites, 1987, 8, 308.

10. P.J. Flory, 'Principles of Polymer Chemistry', Cornell University Press, London, 1971

11. M. Claybourn and M. Reading, Journal of Applied Polymer Science, In Press.

12. M. Bahra and M. Reading, Unpublished.

13. S.V. Wolfe and D.A. Tod, Journal of Macromolecular Science, 1989, A26(1), 249.

14. R.E. Wetton, R.D.L. Marsh, G.M. Foster and G. Connolly, 'ANTEC 91', Montreal, 1991

15. J.C. Duncan, S. O'Donohue and R.E. Wetton, 8th International Conference on Deformation Yield and Fracture of Polymers, 1991, 99/1

16. M. Reading,'Constant Rate Thermal Analysis and Beyond', This Publication.

Controlled Rate Thermal Analysis and Beyond

M. Reading

ICI PAINTS, RESEARCH DEPARTMENT, WEXHAM ROAD, SLOUGH SL2 5DS, BERKS,

1 INTRODUCTION

In the early 1960s an individual, J Rouquerol[1], and a team consisting of two brothers, F Paulik and J Paulik[2], independently arrived at the same idea for a different approach to Thermal Analysis. Conventional Thermal Analysis (TA) often consists of holding the temperature of a sample at a constant value while measuring its response with some sort of transducer. Why not, they reasoned, start with the intention of holding the measured response of the sample to a constant value, eg. rate of mass loss, by changing the temperature in any way necessary to achieve this objective. Rouquerol has compared this general approach with conventional thermal analysis by use of the schematic diagrams shown in Figs. 1 and 2. With conventional TA the temperature of the sample follows some predetermined path as a function of time. With this new approach it is some parameter of the sample that follows a predetermined path as a function of time, this end being achieved by means of changing the sample temperature. Rouquerol[3] has defined this method as "a general thermoanalytical method where a physical or chemical property "X" of a substance is modified, <u>following a pre-determined programme X = f(t),</u> under the appropriate action of temperature".

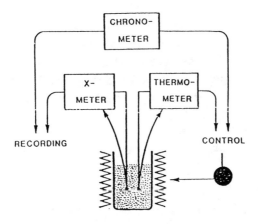

Figure 1 Schematic Representation of Conventional Thermal
Analysis (from ref.3).

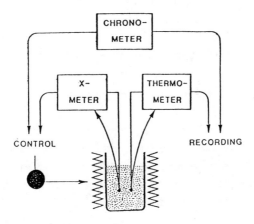

Figure 2 Schematic Representation of Controlled Transformation
Rate Thermal Analysis (from ref.3).

At first sight this might appear like a very complex undertaking with no clear prospect of any benefit. However, the very real benefits that can derive from adopting this approach are the subject of this article. Moreover, there is now the possibility of a whole new family of approaches to temperature programming that, while not completely replacing the conventional approaches, may well become the preferred method in most cases.

2 THE BASIC PRINCIPLES

Introduction

The following discussion is mainly confined to considering the study of solid state decomposition reactions because CRTA has been used predominantly to study reactions of this type. This is in part because this method lends itself most easily to Thermogravimetry and Evolved Gas Analysis, as we shall see below, but also because it is with this type of reaction that the advantages of this new approach over conventional methods are greatest. Nevertheless the basic principles described below can be adopted in any form of Thermal Analysis. This point will be further considered below.

The Quasi-Isothermal Quasi-Isobaric method

The Pauliks based their apparatus on a thermogravimetric balance. They derived a control system that would heat a sample in such a way as to maintain the rate of mass loss at a constant value. The measured parameter became the temperature as a function of time. Their apparatus is illustrated schematically in Fig.6. This type of experiment often produces temperature traces that were almost isothermal during much of the reaction hence the name Quasi-isothermal[4]. The Pauliks proposed the abbreviation Q-TG for this method. They also designed a number of crucibles from which the escape of evolved gases to the outside atmosphere was restricted in such a way as to

force within the crucible the rapid generation of an
atmosphere consisting almost entirely of the evolved species
at the ambient pressure[4]. When experiments were carried out
under these conditions the Pauliks used the term Quasi-
Isothermal, Quasi-Isobaric. The type of results they
obtained are illustrated in Fig.3.

Constant Rate Thermal Analysis

Rouquerol's approach relied upon using a transducer to
monitor the pressure of evolved gas in a continuously
evacuated chamber. He designed a controller that heated the
sample in such a way as to maintain the monitored pressure
constant. As the pressure is maintained constant the rate at
which the gas is pumped away remains constant thus the rate
of mass loss, when a single gas is evolved, is also
constant. Subsequently he coupled this type of approach with
vacuum thermogravimetry[5], and this apparatus is illustrated
schematically in Fig.7. The type of results he obtained are
illustrated in Fig.4.

The similarity between these techniques is immediately
apparent. Both maintain the reaction rate constant, both
control the pressure of the evolved species in the reaction
environment. The differences lie simply in the means adopted
to achieve these ends and the pressure regime they are best
suited for. The name Controlled Transformation Rate Thermal
Analysis has been proposed by Rouquerol[3] to describe this
approach in which the heating programme is determined by the
reaction(s) taking place rather than by a predetermined
heating programme. The word Transformation can
be dropped when it is clear that the controlled parameter is
not the heating rate but the rate of change of a measured
property of the sample. The word Constant can be substituted
for Controlled when this is case. The most frequently used
abbreviation is CRTA. This nomenclature is supported by the
ICTA Nomenclature Committee and is certainly, in this
author's view, preferable to that proposed by the Pauliks.

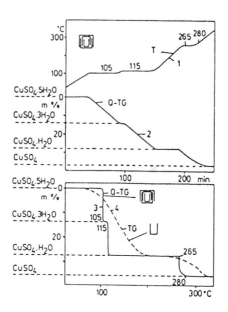

<u>Figure 3</u> Results from Quasi-Isothermal Quasi-Isobaric Thermal
Analysis (Q-TG) for Copper Sulphate Pentahydrate.
Temperature (1) and Mass (2) Against Time for Q-TG.
Mass from Q-TG (3) and Mass From Linear Rising
Temperature (4) Against Temperature (from ref.4).

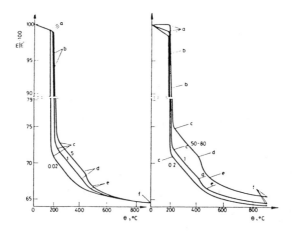

<u>Figure 4</u> Results from Constant Rate Thermal Analysis (CRTA) for
Gibbsite. Left Shows the Influence of Pressure Given in
Torr. Right Shows the Influence of Grain Size in Microns
(from ref.14).

3 THE THEORETICAL ADVANTAGES OF CRTA

A comparison between CRTA and conventional isothermal and rising temperature experiments is made in Fig. 5 in terms of a typical thermogravimetric run coupled with Evolved Gas Analysis. One advantage is that CRTA enables a more precise control of the uniformity of the reaction environment which broadly consists of controlling the product gas pressure and the temperature and pressure gradients within the sample bed. Inspection of the results given in Fig. 5 clearly shows the advantage of CRTA in terms of controlling the product gas pressure which this method maintains constant.

The problem of pressure and temperature gradients within the sample bed is usually addressed by using small samples sizes, thus reducing mass and thermal transport problems. However, it should be remembered that another important factor is reaction rate; the higher the reaction rate of an endothermic decomposition the greater the chance of significant temperature and pressure gradients within the sample. This poses special problems for isothermal experiments because the dangers of having too high a reaction rate at the beginning of an experiment must be balanced against the problems of too slow a reaction rate near the end thus leading to incomplete reaction over realistic time scales. Linear rising temperature experiments pose a similar problem. With this type of experiment the rate goes through a maximum near the mid point of a reaction step. Adjusting the heating rate to limit this maximum can also give rise to unrealistically long experiments. Once a rate is chosen that is sufficiently low to avoid excessive temperature gradients, CRTA ensures that the experiment will take the minimum possible time compatible with obtaining undistorted results. In this light it can be seen that CRTA is an eminently practical approach.

Because it enables much better control of the sample environment, making it possible for uniform conditions to exist throughout the sample bed, it is a very good

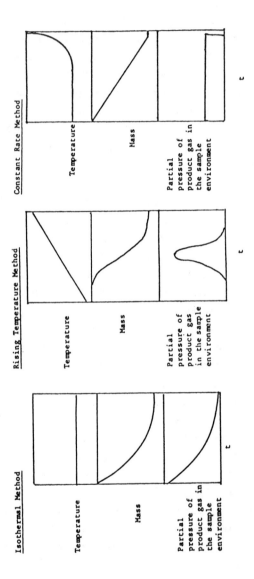

Figure 5 Comparison of the Results From Different Methods (from ref. 11).

preparative method for porous or finely divided solids that
are prepared by thermal decomposition. For the same reasons
CRTA avoids artifacts that stem from having a non-uniform
sample environment, such as artificial broadening of the
temperature interval over which a reaction occurs, and is
consequently also to be preferred for measuring kinetic
parameters.

A common feature of heterogeneous decomposition
reactions is that the reaction rate can be controlled by the
rate of diffusion of the gaseous product from the reaction
interface because it finds its access to the external
environment restricted by the intervening layer of already
reacted material. Whether the diffusion step is the rate
limiting one will generally depend upon the rate of gas
evolution. As Stacey has pointed out[6], by controlling
reaction rate CRTA offers the possibility of moving in a
controlled manner from a diffusion controlled regime to one
where the interface geometry is the governing factor.

Another advantage of CRTA is that it improves
resolution. Using a linear temperature ramp there is a
gradual build up of reaction rate as the reaction starts
then a gradual tail off as it finishes. In contrast the
start of a reaction under CRTA is often marked by an abrupt
change from a steeply rising temperature to an almost
isothermal plateau (this depends upon the reaction mechanism
but this behaviour is frequently encountered, see ref. 4).
The end of the reaction is similarly well delineated as the
temperature rises exponentially. Furthermore, by effectively
reducing the heating rate as the reaction rate accelerates,
the temperature window within which the decomposition occurs
is narrowed. For both of these reasons the ability of CRTA
to clearly resolve the different steps in multi-step
processes is generally much better than conventional
methods.

In summary, it can be seen that CRTA provides a number
of advantages. Because it enables a greater control over the
reaction environment it provides;

1) Results that are less liable to contain artifacts due to inhomogeneous reaction throughout the sample bed.

2) The ability to pass from a diffusion controlled regime to one determined by the geometry of the reaction interface in a controlled manner.

3) Better kinetic data.

4) Better defined preparative conditions for porous or finely divided solids.

Because it allows reactions to occur in a much narrower temperature window and clearly delineates the beginning and end of a reaction it provides;

5) Better resolution for a given duration of experiment.

4 EXISTING FORMS OF CRTA

Two of the existing forms are given in Figs. 6 and 7. One of the advantages of the method adopted by Rouquerol, which originally consisted simply of a pumping system and a pressure transducer, is that this method can easily be coupled with other techniques. Fig.7 shows how it can be coupled to a thermobalance[5], it has also been coupled with microcalorimetry[7]. In a further development a mass spectrometer was substituted for the pressure gauge[8] thus making possible the control of a single mass peak and therefore, under favourable conditions, the control of a single evolved species.

Stacey developed the first apparatus based on controlling the partial pressure of water using a dew point hygrometer[9]. This was used in series with a katharometer so the information could be gained about the rate of evolution of other species, see Fig.8. This apparatus was used to study thermal decompositions and catalyst activation by using reactive (hydrogen containing) atmospheres.

The present author developed the first CRTA apparatus based on the use of infra-red detectors[10] as shown in

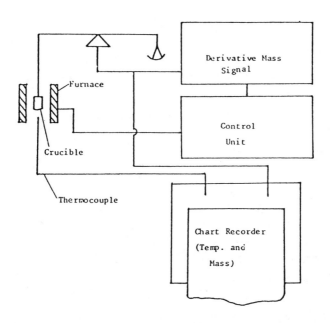

Figure 6 Schematic Diagram of the Paulik's Apparatus for Q-TG.

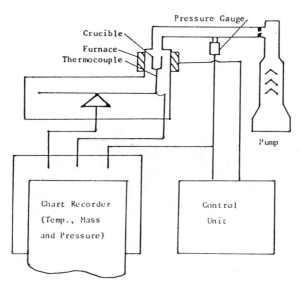

Figure 7 Schematic Diagram of Rouquerol's Apparatus for CRTA Coupled to Thermogravimetry.

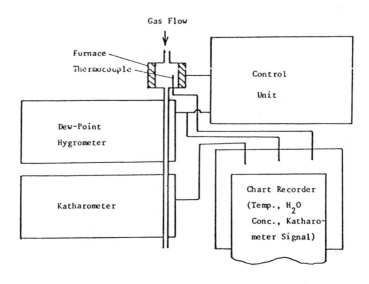

Figure 8 Schematic Diagram of Stacey's Apparatus for CRTA Under Flowing Atmospheres.

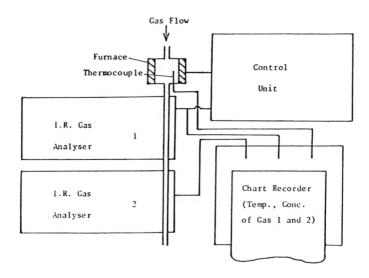

Figure 9 Schematic Diagram of Reading's Apparatus for CRTA Under Flowing Atmospheres.

Fig.9. This instrument had the advantage that it enabled the control of the rate of evolution of any infra-red active gas, rather than just water, while simultaneously monitoring the evolution of another. It was also developed for the study of catalyst preparation and reactivity.

An article published by this author in 1988[11] finished with an appeal to Instrument manufacturers to make a commercially available CRTA apparatus. Recently TA Instruments has begun offering as an option on their TGA2950 an implementation similar to that of the Paulik's. It is unlikely, however, that these two events are connected. Nevertheless it is now probable that other manufacturers will follow suit and Constant Rate TGA will become widely available.

Although the discussion so far has been confined to decomposition reactions followed by TGA and EGA, there is no reason in principle why this technique could not be applied to any thermal method measuring any property. This point has been well made by Rouquerol who has reviewed the different applications to date[3]. A further illustration of this is given below.

5 APPLICATIONS OF CRTA

Kinetics

The kinetics of solid state reactions are usually taken to be of the form;

$$d\alpha/dt = f(\alpha)Ae^{-E/RT} \qquad - 1$$

Where α = fractional extent of conversion

           $f(\alpha)$= some function of extent of conversion

           A   = the pre-exponential constant

           E   = the activation energy

           R   = the gas constant

           T   = the absolute temperature

The parameters that must be defined are $f(\alpha)$, A and E. Obtaining agreement between different workers for nominally the same reaction is notoriously difficult for heterogeneous solid state reactions[12]. It is possible that CRTA will provide the tool that will overcome this problem.

That the theoretical advantages of CRTA can translate into real practical advantages has been adequately demonstrated[11,12,13]. A particularly powerful means of measuring activation energies is the rate jump method illustrated in Fig.10 for calcium carbonate. The reaction rate jumps between two preset values and the corresponding temperature jump is measured. The temperature immediately before and after the rate jump are taken together with the two different rates and it is assumed that $f(\alpha)$ does not change significantly during the jump itself. From equation 1 it can easily be shown that;

$$E = -\frac{R \ln[(d\alpha/dt)_1/(d\alpha/dt)_2]}{(T_2-T_1)/(T_1T_2)} \qquad - 2$$

Where E = the activation energy. The other variables can be taken from Fig.10. In this case only values from the low to high jump were used as they exhibited the most well controlled behaviour. The activation energy obtained was 193 KJ/mol which is in good agreement with the previously determined value for this compound[12].

A further advantage of the CRTA method is the possibility of using the reduced temperature plot method of determining $f(\alpha)$. Table 1 summarises the most commonly used kinetic expressions, it should be noted that they fall into three categories. The first includes models that assume the formation of sparse nuclei that grow, resulting in an accelerating reaction rate, then merge. These expressions are known as Avrami Erofe'ev equations (numbers 1–3 in table 1). The second assumes that the surface is rapidly covered with diffuse overlapping nuclei then the reaction interface proceeds through the sample particle. These are known as order or geometric expressions (numbers 4-6 in table 1).

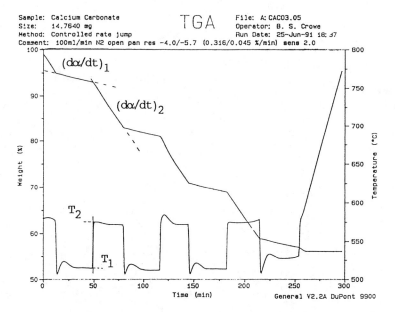

Sample: Calcium Carbonate
Size: 14.7640 mg
Method: Controlled rate jump
Comment: 100ml/min N2 open pan res -4.0/-5.7 (0.316/0.045 %/min) sens 2.0

File: A: CACO3.05
Operator: B. S. Crowe
Run Date: 25-Jun-91 18: 37

Figure 10 The Rate Jump Method Applied to Calcium Carbonate.

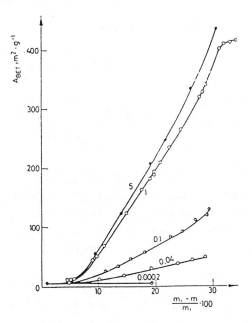

Figure 11 Effect of Pressure on the Development of Surface Area
During the Decomposition of Gibbsite (from ref.14).

No.		$f(\alpha)$ $[=(d\alpha/dt)/k]$
	Sigmoid Rate Equations	
1		$(1-\alpha)(-\ln(1-\alpha))^{\frac{1}{2}}$
2	Avrami Erofe'ev	$(1-\alpha)(-\ln(1-\alpha))^{\frac{2}{3}}$
3		$(1-\alpha)(-\ln(1-\alpha))^{\frac{3}{4}}$
	Deceleratory	
4	first order	$(1-\alpha)$
	Based on Geometric Models	
5	Contracting area	$(1-\alpha)^{\frac{1}{2}}$
6	Contracting volume	$(1-\alpha)^{\frac{2}{3}}$
	Based on Diffusion Mechanism	
7	One dimensional diffusion	α^{-1}
8	Two dimensional diffusion	$[-\ln(1-\alpha)]^{-1}$
9	Three dimensional diffusion	$[1-(1-\alpha)^{\frac{1}{3}}]^{-1}(1-\alpha)^{\frac{2}{3}}$
10	Ginstling Brounshtein	$[(1-\alpha)^{-\frac{1}{3}}-1]^{-1}$

Table 1 Commonly Used $f(\alpha)$

The third category assumes diffusion to be the rate limiting process.

Fig.12 shows the reduced temperature plots for these kinetic models (the details of this method can be found in ref. 11). While it is not possible to distinguish between different order expressions using this method it does provide a means of distinguishing between the different categories of mechanism more sensitively than conventional techniques[11]. If a reaction is found to follow an order expression the exact order can be established once the rate jump method has been used to determine the activation energy.

It should be noted that it has recently been shown[13], using conventional kinetic arguments, why the Quasi-Isobaric approach, ie. CRTA with high product gas pressures, or any approach that allows the build up of high concentrations of the gaseous product, can not provide good kinetic data.

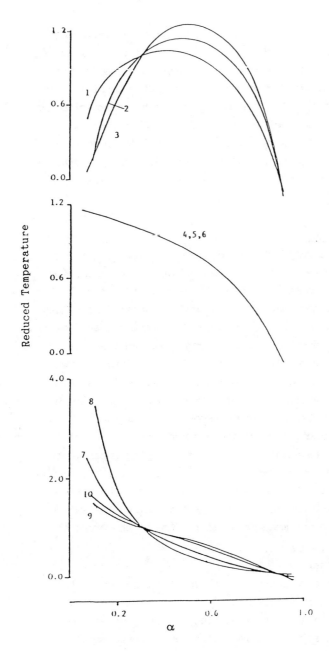

Figure 12 Reduced Temperature Master Plots (from ref.11).
Numbers Correspond to the Equations in Table 1.

This fact goes a long way toward explaining the differences
of opinion that have existed about the validity of kinetics
measured using CRTA and other non-isothermal methods. There
is now a good body of evidence obtained using calcium
carbonate[11,12,13] which, taken together with the results
presented here, strongly supports the argument that
conventional Arrhenius kinetics can be applied to solid
state reactions and that CRTA offers significant advantages
in this field of study when dealing with decomposition
reactions.

Preparative Methods

Rouquerol illustrated the profound effect product gas
pressure can have on the surface area of the decomposition
products of gibbsite[14]. Fig.11. illustrates how changing the
pressure above the sample from 0.0002 to 5 Torr (the values
are given next to the curves in Torr) changes the surface
area of the product by about two orders of magnitude at any
point in the decomposition. While this was already known in
general terms, Rouquerol's work illustrates well how CRTA
can be used to exert a much greater degree of control over
the reaction environment and consequently over the
properties of the decomposition product.

Stacey revealed a similar effect with his experiments
on the preparation of catalysts[9]. He found that changing the
partial pressure of water when decomposing zinc hydroxide
from 658 to 2150 Pa had a significant affect on the surface
area, pore size and mean crystallite size of the product.

Enhanced Resolution

This point has been dealt with above. The enhancement that
can be achieved is illustrated by Fig.3. Perhaps all that
needs to be said under this heading is that, for a given
duration of experiment, CRTA will almost always produce
better resolution than a linear heating programme[1,4].

6 ALTERNATIVE APPROACHES

Introduction

Another technique that is similar to the ones described above has been proposed by Sorensen called Stepwise Isothermal[15] Analysis. In this method an upper and lower limit is set for the reaction rate. When it falls below the minimum the temperature is increased until it exceeds the maximum then the temperature is held constant until the reaction rate again falls below the minimum. This method has been applied to thermogravimetry and to dilatometry.

In addition to this a method has also recently been developed by TA Instruments called Dynamic Rate TGA[16] in which the heating rate is dynamically and continuously varied in response to the sample's rate of mass loss.

The interesting point about both of these methods is that they do not fall within the definition of either conventional thermal analysis or CRTA as illustrated by Figs. 1 and 2. They are more accurately represented by Fig.13. The power delivered to the furnace by the controller at any point in time is a function both of the measured response of the sample and of the rate of change of the sample temperature. When the entire course of the experiment is considered it is not the intention of these methods to maintain anything at a constant value, rather they adopt a control strategy designed to constrain or limit not only the reaction rate but also the heating rate, within certain limits.

This poses a problem of nomenclature. We propose the term Constrained Rate Thermal Analysis. The different forms could be distinguished between using a single lower case letter, ie. CnRTA for ConstantRTA, CrRTA for ControlledRTA and CaRTA for ConstrainedRTA. Another possibility is Optimising Temperature Programme Thermal Analysis. This name is meant to imply that the objective in following a particular temperature programme is not to maintain any

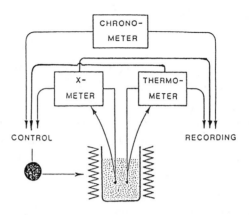

Figure 13 Schematic Representation of Constrained Rate Thermal Analysis.

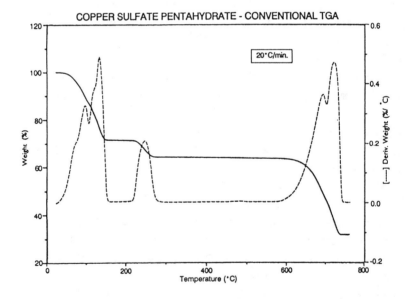

Figure 14 Results From a Linear Rising Temperature Experiment on Copper Sulphate Pentahydrate (from ref.16)

quantity constant, or even to control it in the sense of making it follow a predetermined path as a function of time. Rather, it is to optimise the data gathered by the experiment in terms of, for example, achieving maximum resolution in a given time.

In this author's opinion, the step-wise isothermal method has the very undesirable feature that, depending on the values of the upper and lower limits, a single reaction step could appear as a number of discrete steps thereby presenting a misleading picture. When the limits are set very close together then it becomes very similar to CnRTA. For these reasons this method will not be considered further here. The Dynamic Rate method, however, presents the possibility of having a smooth continuous response that is different in principle to CRTA.

Possible Advantages of Constrained Rate or Optimising Temperature Control Thermal Analysis

Why might this approach be preferable to CrRTA ? One possibility is that CrRTA is difficult to achieve in practice for high reaction rates which are desirable in industrial laboratories to achieve a fast throughput of samples. The Dynamic Rate method can offer a workable approach that improves resolution without sacrificing speed. Fig.s 14 and 15 demonstrate that significant improvements in resolution can be achieved yet both experiments took a similar amount of time to cover the same temperature range[16]. While possible in principle, it is difficult, in this author's experience, in practice to perform properly controlled CnRTA at speeds that would give results in a comparable time.

A more fundamental point is that, for a given length of experiment, CrRTA may not be the best way of maximising resolution. If maximum resolution is the outcome being sought, then concerns about maintaining the evolved gas at a constant pressure are secondary (or irrelevant if the

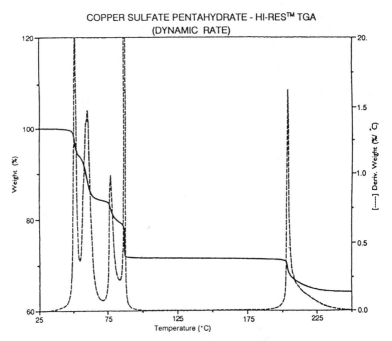

Figure 15 Results From a Dynamic Rate Experiment on Copper
Sulphate Pentahydrate (from ref.16)

High Resolution TGA

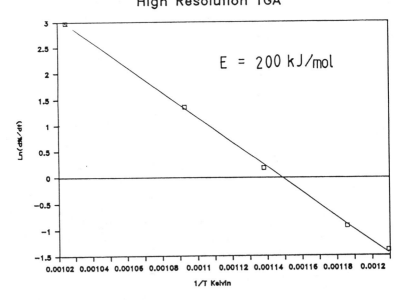

Figure 16 Arrhenius Plot for Calcium Carbonate From Dynamic Rate
Results.

process being considered does not involve the evolution of a gas). If a two stage reaction is considered then maximum time, ie. slowest reaction rate, should be spent during the transition from one stage to the other, before and after this point the reaction could proceed faster. To determine whether a transition between two reaction steps is occurring the rate of change of temperature must be taken into account, simply monitoring sample response without sample temperature could not produce the desired result. Such a system must conform to the schematic shown in Fig.13 and thus fall outside of the current definition of CrRTA but within the proposed definition of CaRTA or Optimising Temperature Programme Thermal Analysis.

Modern microprocessor based controllers mean that very complex strategies can be adopted to achieve specific results for a given type of system. There is no absolute requirement that any single parameter be held at a constant value especially in homogeneous systems. The Dynamic Rate method may be the first of a number of approaches to exploit these new possibilities for 'intelligent' temperature control.

Some Kinetic Results

It is interesting to consider the use of the Dynamic Rate method for looking at kinetics. A series of experiments were carried out on the TA Instruments TGA2956 at different 'resolution values' (effectively lower and higher average reaction rates). The log of the rate of reaction at the mid-point of the reaction ($\alpha=0.5$) are plotted against temperature in Fig.16. From equation 1 it can be seen that the slope of this curve is -E/R. The value for E obtained was 200 kJ/mol which is in excellent agreement with the value obtained from the rate jump experiment above and with our previous work[12]. While CnRTA is to be preferred for kinetic measurements the Dynamic Rate method can give acceptable results.

It is also possible to modify the reduced temperature method to deal with non-constant rate experiments. From equation 1, by following a derivation parallel to that for the original reduced temperature method[11], it can be shown that

$$
\frac{(1/T_\alpha - 1/T_{0.8})}{(1/T_{0.2} - 1/T_{0.8})} =
$$

$$
\frac{\ln[\,(f(\alpha)/f(0.8))/((d\alpha/dt)_\alpha/(d\alpha/dt)_{0.8})\,]}{\ln[\,(f(0.2)/f(0.8))/((d\alpha/dt)_{0.2}/(d\alpha/dt)_{0.8})\,]}
$$

$$- 3$$

Where T_α = The temperature at given value of α

$T_{0.2}$ = The temperature at $\alpha=0.2$ (similarly at $\alpha=0.8$)

$f(\alpha)$ = The alpha function at a given value of α

$f(0.2)$ = The alpha function at $\alpha=0.2$ (similarly at $\alpha=0.8$)

While this formula cannot be used to generate true master curves, it can be used to generate 'corrected' master curve values for comparison with the experimental reduced temperature data. The left hand side of the above equation (the Reduced Temperature) is plotted against values for the right hand side generated using the different possible alpha functions. A straight line with a slope of one that passes through the origin signifies that the expression being considered is the correct one. Fig.s 17, 18 and 19 show the results for one of the mid-range Dynamic Rate results. The numbers given in the figures correspond to those for the equations in table 1. It can be seen the type of expressions that give the most satisfactory results are the order equations. A precise order could be calculated but this is

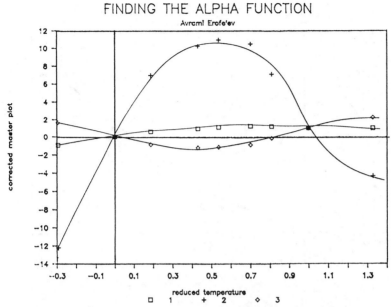

<u>Figure 17</u> Plots of Corrected Master Plot Against Reduced
Temperature for Calcium Carbonate From Dynamic
Rate Results.

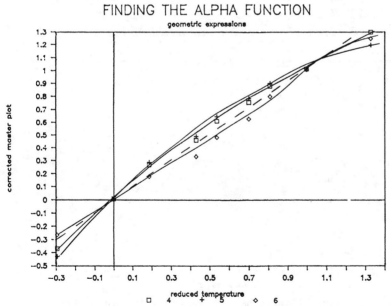

<u>Figure 18</u> Plots of Corrected Master Plot Against Reduced
Temperature for Calcium Carbonate From Dynamic
Rate Results. Dotted Line Shows Perfect Fit.

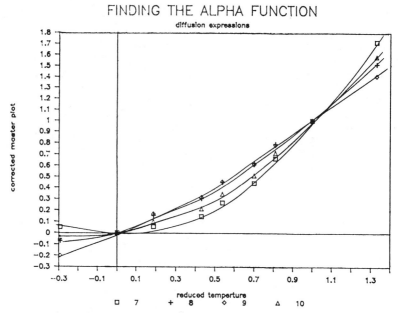

Figure 19   Plots of Corrected Master Plot Against Reduced
              Temperature for Calcium Carbonate From Dynamic
              Rate Results.

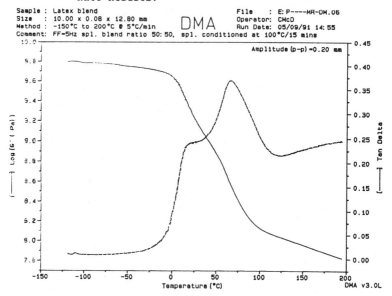

Figure 20   Conventional DMA Results for a Mixture of Two
              Different Polymers.

not attempted here. The general conclusion that this
reaction conforms to an order expression is in good
agreement with previous results[13]. We have shown that the
determination of the exact order of this reaction is
difficult and may even be dependent upon the experimental
conditions, but there is good evidence that this reaction
conforms to this class of behaviour. Consequently the
conclusion drawn from the dynamic rate data is in good
agreement with that drawn from CnRTA and other results.

It is perhaps worth noting that this method of analysis
can be applied to other temperature programmes such as a
conventional linear ramp.

Other Possible Forms of Constrained Rate Thermal Analysis

A good example of a technique that could benefit from
a CnRTA or CaRTA approach is DMA. Fig.20 shows a
conventional DMA experiment performed on a mixture of two
different latexes with different Tg's. The result is
adequate to show that there are two phases in the system,
however, the first transition appears only as a shoulder on
the second. Figs. 21 and 22 show what this author believes
is the first example of a DMA experiment carried out under
CaRTA conditions. The experiment takes no longer than the
conventional one but the resolution is much better. When
dealing with transitions of this type that are
quasi-reversible and do not involve a gaseous product there
is no special advantage to maintaining the rate at which the
transition takes place at a constant value. Furthermore,
there is now potential for the further extension of the
constrained rate approach to take into account both the
storage and loss moduli at the same time viz;

$$dT/dt = f(t, T, G', G'') \qquad\qquad - 4$$

One strategy could be to prevent the rate of change of
any single measured variable exceeding certain preset values
with the subsidiary constraint that they would not be

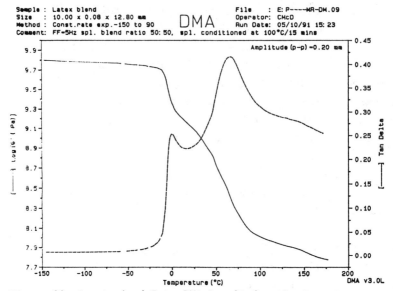

Figure 21 Constrained Rate DMA Result for the Same System as in Figure 20 Plotted Against Temperature.

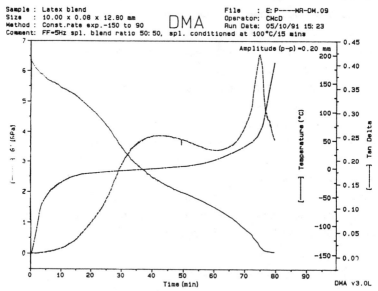

Figure 22 Constrained Rate DMA Result for the Same System as in Figure 20 Plotted Against Time.

allowed to fall below preset values until a preset maximum
temperature is achieved. The challenge is to devise the most
efficient algorithms for maximising, for example,
resolutions in the minimum experimental time. Intelligent
optimising algorithms could be devised which begin with a
set of constraints chosen by the operator but which can be
changed as a function of the response of the sample. A full
discussion of this topic is beyond the scope of this
article. However, this author believes that a new range of
possibilities for more sophisticated approaches to
temperature programming is now opened up by the use of
microprocessor based controllers reacting intelligently to
the observed sample response(s) as a function of time and
temperature.

7 CONCLUSIONS

We have seen that Controlled Rate Thermal Analysis
represents a new approach for thermal methods that offers
significant advantages over conventional methods. We have
also seen that there are other possible approaches where the
temperature programme the sample is subjected to is not
simply determined by the need for one measured chemical or
physical property of a sample to follow a predetermined path
as a function of time. For this reason they fall outside of
the current definition of CRTA. We have proposed a possible
scheme of nomenclature and suggested that this alternative
approach, where a number of different factors can be taken
into consideration, be called Constrained Rate Thermal
Analysis (CaRTA) or Optimising Temperature Programme Thermal
Analysis. Results generated by Dynamic (heating) Rate TGA, a
recently introduced form of CaRTA, have been presented and
shown to give good resolution and kinetic results. The
possible advantages of Applying CaRTA to DMA have been
illustrated and discussed.

Conventional methods of temperature programming can be described as programme determined temperature control, where the temperature programme the sample is subjected to is determined in advance by the experimenter and proceeds independent of the response, if any, of the sample. The alternative approaches discussed above could be described as sample determined temperature control where the temperature programmed the sample is subjected to is largely determined by the sample response within constraints determined by the experimenter. To date the most widely used form of sample determined temperature control has been Constant Rate Thermal Analysis and this can be seen to have numerous advantages compared to more conventional methods. Now alternatives to CnTRA are appearing that might offer advantages of their own. Whatever the details of the technique used it is becoming increasingly obvious that a move from programme determined temperature control to more and more sophisticated methods of sample determined temperature control will significantly improve the performance of all of the currently familiar techniques of thermal analysis. This change is to be welcomed and both the users and the manufactures of thermoanalytical instruments will benefit from exploring and promoting the interesting new opportunities that are in store.

Acknowledgements

The author would like to thank David Elliott for reading this manuscript and making helpful suggestions and Cathy McDermott and Malkit Bahra for carrying out the experimental work. Thanks are also due to Marlin Scientific and TA Instruments, in particular Ben Crowe, for providing access to their Dynamic Rate and CRTA instrument. Finally the author would like to thank those publishers who kindly gave their permission for material to be reprinted.

<u>REFERENCES</u>

1) J Rouquerol, <u>Bull. Soc. Chim. Fr.</u>, 1964, <u>31</u>

2) a) L. Erdey, F. Paulik and J. Paulik, Hungarian Patent
   No. 152197, 1962
   b) J. Paulik and F. Paulik, <u>Anal. Chim. Acta</u>, 1971, <u>56</u>

3) J Rouquerol, <u>Thermochim. Acta</u>, 1989, <u>144</u>, 209

4) F. Paulik and J. Paulik, <u>Thermochim. Acta</u>, 1986, <u>100</u>, 23

5) J. Rouquerol, <u>J. Therm. Anal.</u>, 1970, <u>2</u>

6) M.H. Stacey, <u>Anal. Proc.</u>, 1985, <u>22</u>, 242

7) M. Ganteaume and J. Rouquerol, <u>J. Therm. Anal.</u>, 1971, <u>3</u>,
   413

8) G. Thevand, F. Rouquerol and J. Rouquerol, Thermal
   Analysis, B. Miller Editor, John Wiley and Sons, New
   York, 1982, Vol.2, 1524

9) M.H. Stacey, Proceedings of the 2nd ESTA, D Dollimore
   Editor, Heydon, London, 1981.

10) M. Reading and J. Rouquerol, <u>Thermochim. Acta</u>, 1985, <u>85</u>,
    299

11) M. Reading, <u>Thermochim. Acta</u>, 1988, <u>135</u>, 37

12) M. Reading, D. Dollimore, J. Rouquerol and F. Rouquerol,
    <u>J. Therm. Anal.</u>, 1984, <u>29</u>, 775

13) M. Reading, D. Dollimore and R. Whitehead, <u>J. Therm.
    Anal.</u>, in press

14) J. Rouquerol and M. Ganteaume, <u>J. Therm. Anal.</u>,1977, <u>11</u>,
    201

15) O. Toft Sorensen, <u>Thermochim. Acata.</u>,1981, <u>50</u>, 163

16) S.R. Sauerbrunn, P.S. Gill and B.S. Crowe, 5th ESTAC,
    1991, O-6 (Proceedings to be published in J. Therm.
    Anal.)

Applications of Thermal Analysis of Polymers

J. N. Hay

THE SCHOOL OF CHEMISTRY, THE UNIVERSITY OF BIRMINGHAM, BIRMINGHAM
B15 2TT, UK

1 INTRODUCTION

The last few decades have seen the proliferation of thermal analytical techniques and these have become very widely applied to the characterisation of thermoplastic materials, such that most research papers on the structure/property relationships refer to one or more of these techniques. Polymers are a major area of application, and several industrial standard procedures for characterising the extent of cure, or measurement of a degree of crystallinity involve these techniques routinely. Polymers invariably do not require extension to the analytical techniques, indeed since they are mainly organic based molecules a lower temperature range, 150–800K is usually sufficient. Although thermal analysis covers a diverse range of techniques it will not be possible to cover more than a few of the more popular methods and to concentrate on a wider range of applications. However, it will become increasingly apparent that certain techniques complement one another and together are more powerful than one on its own. Whether these are separate or multiple probe instruments is a matter of personal choice and circumstances.

Modern thermal analytical techniques are very powerful, data is readily accumulated and processed by data stations. However,the interpretation of the data must be made with a full apprecation of the behaviour of polymer materials. In particular, polymers are visco-elastic, in that they have the characteristics of an elastic solid and a viscous liquid. Properties are temperature and time dependent, such that thermal history is an important variable. This imposes certain restrictions in any study of the effect of molecular structure on properties and physical transitions, in general.

These effects will be discussed in some detail, in the application of differential scanning calorimetry, DSC to the study of physical properties of polymers.

2   EXPERIMENTAL

A Perkin-Elmer Differential Scanning Calorimeter, DSC-2, was used interfaced to a BBC Master microcomputer. This collected energy flow rate data as a function of temperature and time. Typically 1120 data points were stored for each analysis and starting and finishing isotherms were also stored. Heat, cool and hold functions were controlled by the computer and routine analysis carried out automatically by the computer software. The data was directly stored on disc for later processing.

Flat discs, 3mm diameter, were cut from moulded polymer sheets, weighed to 0.01mg and placed in the aluminium sample pans. Lids were used to prevent contamination of the DSC sample holder. Prior to this the aluminium pan and reference pan were separately measured on the DSC and the data obtained used as an instrument baseline. The energy response of the DSC was calibrated using the enthalpy of fusion of indium, 28.45 Jg^{-1}, and the temperature calibrated from the m.pts. of ultra-pure metals, indium, tin and lead. Corrections were made for thermal lags.

A Dynamic Mechanical Thermal Analyser, DMTA, from Polymer Laboratories Ltd. was used interfaced to an IBM microprocessor, System 2 model 30. Polymer bar samples were clamped and used in a flexing mode. Measurements of the flexural modulus, E' and tanδ were made from 120 to

500K at a heating rate of 2-3K min^{-1} and over the frequency range 0.01 to 100Hz with mechanical displacements between 0.01 and 0.25mm.

Similarly a Dielectric Thermal Analyser, DETA, manufactured by Polymer Laboratories Ltd. was used to measure the dielectric response of disc samples placed between electrodes. The frequency range was 20 to 10Hz. Both instruments could carry out multifrequency scans from 120 to 600K at heating rates between 0.5 to 20 Kmin^{-1}.

3 DIFFERENTIAL SCANNING CALORIMETRY

DSC is defined as a technique in which "the difference in energy inputs into a sample and reference is measured as a function of temperature, or time, while they are subjected to a controlled temperature programme". A power compensated DSC was used and the rate of energy absorbed by the sample plotted against temperature as the sample was heated at a constant rate. A typical thermal analysis for a crystallisable amorphous polymer is shown in Figure 1, in which heat flow is plotted against temperature. Correction has been made for the aluminium sample pans and the response reflects the heat capacity change of the polymer with temperature in the absence of a chemical or physical change. The rate of heat absorbed,

$$w = dH/dt = dH/dT \cdot dT/dt = C_p \cdot R \tag{1}$$

where H is the enthalpy of the polymer, t is time and T is temperature. C_p is the heat capacity per g, and R the rate of heating.

Accordingly the deflection from the instrument baseline is a measure of the specific heat of the sample. If the sample undergoes a physical or chemical change there will be a corresponding change in enthalpy, and a deflection from the specific heat trace as an exo- or endothermic peak. The area under this deflection is a measure of the enthalpy of the change, ΔH,

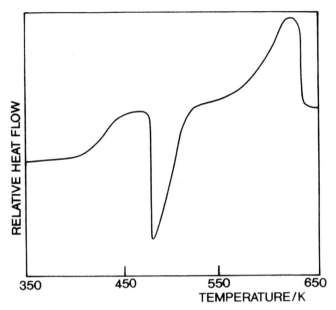

Fig.1   Thermal analysis of an amorphous polymer

$$\Delta H = \Sigma wdt = \Sigma Cp.R.dt \tag{2}$$

and equation 2 is the basis of the thermal analysis of polymers. Figure 1 displays specific heat of a crystallisable amorphous polymer, such as PEEK or PET, as a function of temperature. A low temperature glass transition can be seen as a step increase in the specific heat, followed by a crystallisation exotherm – this we will call low temperature crystallisation – and finally a melting endotherm. At higher temperatures decomposition reactions occur as the polymer decomposes by splitting off volatiles. Studying thermal decomposition in a DSC is not recommended as the volatiles contaminate the platinum sample holders and substantially reduce the lifetime of the calorimeter by decreasing sensitivity and increasing noise levels. On cooling the polymer sample from just above the m.pt. at a slow rate of cooling crystallisation occurs at a much higher temperature than observed on heating the amorphous polymer. This we will call high temperature crystallisation, If this crystalline sample is re-examined on the DSC, see Figure 2, then substantial differences are observed from the original in Figure 1. The glass transition step change in heat capacity is substantially reduced and the low temperature crystallization has disappeared. Two melting peaks instead are observed, the position of the lower reflecting the temperature at which the sample was crystallised. The general shape and positions of the transitions can be used to confirm the structure of a polymer, in that the temperatures of the transitions are characteristic of a polymer, and to a limited extent can be used in a semi-empirical way to characterise polymers, particularly by comparison with known material.

If a low molecular weight solid resin is examined in a similar way, on melting the curing reaction can be followed to apparent completion by the enthalpy change associated with the cure. Cooling and reheating will not reproduce the original melting transition, but usually a glass transition of the cured material only is observed. If the resin is not fully cured further

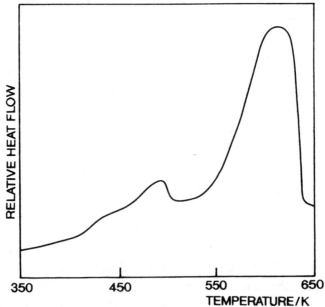

Fig.2   Thermal analysis of crystalline polymer

reaction can occur at the higher temperature, and a subsequent reanalysis will show a change in the glass transition temperature. This is the basis of a standard procedure for defining whether resins have fully cured or not.

DSC is widely used to:

a) measure the glass transition and extent of physical aging in thermoplastic materials,

b) measure crystallization kinetics,

c) assess the degree of crystallinity, and

d) study the melting behaviour and the quality of crystallinity.

We will consider these transitions separately.

4   THE GLASS TRANSITION AND PHYSICAL AGING.

The glass transition is a second order thermodynamic process, since there is no change in enthalpy unlike melting or crystallisation, but only a change in the heat capacity i.e. dH/dt shows a step change and d^2H/dT^2 goes to infinity. Since there is no sharp change in Cp at Tg it is difficult to define in practice. Many convenient but approximate methods will be suggested to you, involving intersection of tangents to the specific heat plots, see Figure 3, but these procedures are wrought with difficulties and can lead to misleading results.

The thermodynamic definition of T_g is

$$H_g(T_g) = H_1(T_g) \tag{3}$$

for which $H^o_g + aT_g + bT_g^2 = H^o_1 + AT_g + BT_g^2$

where $C_{p,1} = A + BT$, and $C_{p,g} = a + bT$

Subscripts 1 and g refer to the liquid and glass states, and superscript o to standard states. From the temperature dependence of the heat capacities of the liquid and glass well away from the glass transition, and by measuring the area under the heat capacity curve from two fixed temperatures, T_1 and T_2 well above and well below the glass transition region, $H_1^o - H_g^o$ can be

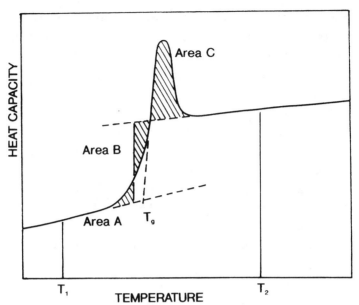

Fig.3 Definition and graphical determination of T_g

eliminated and T_g evaluated. The procedure may appear to
be very involved but we have found it necessary if the
correct trends in the observed value of T_g are to be
observed, see Table 1, where the effect of using an
intersection of tangents procedure is compared with that
of integrating under the specific heat curves. The
expected variation of the measured T_g on the rate of
cooling is totally lost in the former case but apparent
in the latter, such that an activation energy for the
glass forming process can be determined from an
Arrhenius plot of reciprocal temperature of the
transition against logarithm of the cooling rate.

Aras and Richardson[1] have suggested an alternative
method of determining the glass transition temperature,
T_g which also reflects the enthalpic definition but is
more amenable to treatment graphically. The procedure
involves selecting a value for T_g which equates the
areas B and A+C in Figure 3. The procedure is more
sensitive to the determination of T_g since smaller areas
are integrated.

Table 1 Dependence of the measured value of T_g against cooling rate for the glass transition of polyether imide

Cooling Rate/ $K \: min.^{-1}$	Glass Transition Temperature T_g /K	
	Method 1	Method 2 *
160	489.3±0.5	486.3±0.5
80	488.3	485.8
40	487.8	483.8
20	487.3	480.7
10	487.2	479.7
5	487.5	478.8
2.5	488.3	475.2
1.25	488.3	475.1
0.63	489.3	472.8
0.31	488.7	472.5

* Method 1 refers to interpolation of tangents to C_p plots, and Method 2 to integration of C_p plots.

As can be seen from Table 1 the structure of the glass is not unique but is dependent on the rate at which the glass is formed from the liquid, on heat treatment below but close to the glass transition temperature, and any internal stresses or strains. The glass transition is not a characteristic parameter but reflects thermal and mechanical history.[1] To eliminate these effects polymer samples should be cooled from the liquid state at defined rates in order to produce a standardised T_g, but the conditions adopted should be stated in the measurement. The effect of cooling rate or annealing of the glass can give detailed insight into the mechanism of glass formation and the morphology of polymer blends.[2]

Organic glasses are not in equilibrium, and the properties are temperature and time dependent. They gradually approach equilibrium in a temperature range below T_g. The process of physical aging, or more specifically enthalpic relaxation since it is being measured with DSC, is associated with the gradual approach of the non-equilibrium properties of the glass towards the extrapolated equilibrium value characteristic of the liquid. The energetics of physical aging closely follow that of glass formation, and the process is kinetic in origin. If a standard glass, produced from the liquid by cooling at a fixed constant rate, is held for various periods of time at the annealing or aging temperature, T_a, the enthalpy of the glass changes progressively towards the lower extrapolated liquid value at T_a, i.e. $H_l(T_a)$ and on cooling to a lower temperature and reheating to redetermine the glass transition, an endotherm is observed on the transition which increases progressively with annealing time, see Figure 4. The reason for the endotherm is apparent from Figure 5 where the enthalpy of the aged and quenched glasses are compared. On heating the aged glass overshoots the T_g as defined by the interception of the two enthalpy/temperature plots for the aged and liquid states but once sufficient mobility has been achieved the glass relaxes to the liquid dependence with the evolution of the enthalpy change on aging, ΔH_t. The extent of physical aging can be measured by subtracting the area under the specific heat curve of the quenched material from that of the aged material. The maximum amount achieved, ΔH_f, can be seen from Figure 5 to be related to difference between the enthalpy of the quenched and extrapolated liquid at T_a, and in particular

$$\Delta H_f = \Delta C_p . \Delta T \qquad (4)$$

where ΔC_p is the change in heat capacity at T_g of the quenched glass, and $\Delta T = T_g - T_a$. The extent to which aging has been completed is,

$$\Phi(t) = (1-\Delta H_t/H_f) \qquad (5)$$

Plots of $\Phi(t)$ against $\log(t)$ are exponential in shape

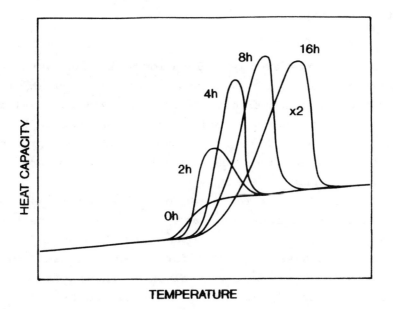

Fig.4   The effect of aging time on the glass transition

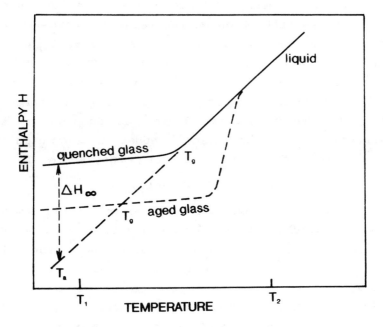

Fig.5   Heat content change on aging

and the kinetics have been interpreted in terms of a stretched exponential,[3]

$$\Phi(t) = \exp(-t/\tau)^{\beta} \tag{6}$$

where τ is the average relaxation time, and β varies between 0 to 1.0. It measures the breadth of the spectrum of relaxation processes involved in physical aging. For homopolymers it varies between 0.3 and 0.6. If $\beta = 1$ there is a single relaxation process involved but as it decreases the breadth of the spectra of relaxations increases.

At long times in excess of τ the aged glass has relaxed to equilibrium and the glass transition temperature, T_g is equal to the aging temperature, T_a. Other methods are available for measuring T_g and it is interesting to compare the values obtained against the time scale of the measuring technique. DMTA and DETA both define T_g from the temperature corresponding to the maximum in tanδ at each frequency, which clearly corresponds to a resonance between the chain mobility and the imposed frequency. A master curve of the dependence of the glass transition on the relaxation time of the imposed frequency is drawn up in Figure 6. This is a composite curve incorporating DETA, DMTA, DSC and physical aging results and covers over about 15 decades in time. Because each technique defines T_g in different ways, the shift factor, log(a) as adopted in the Williams, Landel and Ferry[4] equation was used to superimpose the data at a standard temperature, T_o:

$$\log(a) = -C(T-T_o)/(C_1 + T - T_o) \tag{7}$$

where C and C_1 are constants. The separately determined T_g values overlapped considerably but there is a general progressive increase in the measured glass transition temperature with the frequency of the measurement, DSC measuring the transition in the time scale of seconds and minutes, and physical aging over hours to geological time. The plot indicates that there is a continuum and that physical aging is an extension of the glass-

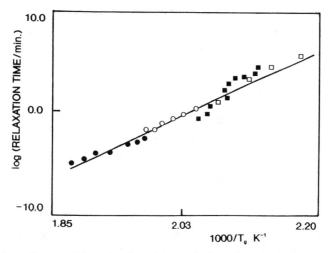

Fig.6 Composite Arrhenius plot of reciprocal glass transition temperature against log (relaxation time) for various techniques.
○DMTA; ●DETA; ■DSC; and □physical aging

forming process over very long time scales. Similar conclusions can be determined from a measurement of the activation energies of the two processes. The general upward curvature of the plot implies a limitation to the temperature range of physical aging, and it has been postulated[5] that as measured by DSC it is limited to the range T_g to T_g-50K.

Physical aging is important in that it is accompanied by a progressive change in material properties with time, and accordingly the yield stress and stiffness increase while fracture toughness and creep characteristics decrease.[6] An evaluation of the size of the effect, the temperature range over which it occurs, and the rate at which it occurs is commercially very important.

5 CRYSTALLISATION KINETICS.

The mechanical properties of polymers are markedly dependent on the degree of crystallinity and it is important to determine the temperature range over which the crystallinity changes and affects material properties.

The fractional extent of crystallisation, X_t, which develops at constant temperature and at time t, is given by the general rate expression, due to Avrami,[7]

$$1 - X_t = \exp -Zt^n \tag{8}$$

where Z is a rate constant which incorporates both crystal growth and nucleation, and n is a constant characteristic of the mechanism. For growth of spherical particles, i.e. spherulitic crystallisation, n should be 3 or 4, but this value is seldom observed due to limitations in the basic kinetic model. Nevertheless, since it is important technologically to be able to define the timescale over which crystallization occurs at various temperatures, to ensure a reasonable degree of crystallinity has developed, the equation has been adopted to describe the overall kinetics.

Two crystallization regimes are observed, although they may not be attainable in all polymers, since the polymer sample cannot be quenched below the temperature of maximum rate without crystallisation occurring. These are :

a) high temperature crystallization close to the melting point, T_m and

b) low temperature crystallization, close to the glass transition temperature.

Isothermal kinetics is conveniently measured by DSC. The polymer sample is first heated to a temperature and for a sufficient time to melt the sample completely but not cause degradation. This can only be considered from the reproducibility of the subsequent crystallisation behaviour. In (a) the calorimeter is then cooled at 180 K/min directly to the crystallisation temperature. The loss of heat from the calorimeter and sample is collected on cooling initially and during subsequent crystallisation until the calorimeter response does not deviate from the isothermal baseline. The crystallisation isotherm is obtained directly by subtracting the cooling curve obtained by cooling the sample through the same temperture difference but at a temperature where crystallisation does not occur, see Figure 7. When this is carried out properly the initial and final baseline are the same.

In case (b) the molten samples are quenched in liquid nitrogen, through the glass transition temperature and the amorphous material heated at 180 K/min. to the crystallisation temperature. The subsequent heating thermogram and crystallisation isotherms are stored on computer. This is also repeated but heating to a temperature at which the sample will not crystallise, and this baseline subtracted from the original as before. The crystallisation isotherms obtained in this way show a marked dependence on crystallisation temperature, see Figure 8a. Both sets of isotherms can be analysed directly since,

$$X_t = \Sigma^t_0(dH/dt).dt/\Sigma^\infty_0(dH/dt).dt \qquad (9)$$

and plots of $\log(-\ln(1-X_t))$ against $\log(t)$ exhibit two linear portions indicative of the presence of two consecutive processes, see Figure 8b. The initial slope, n, is invariably fractional but the second is 1.0. There have been many attempts to modify the Avrami approach to account for these deviations but modified equations appear to be equally poor in describing the overall crystallisation of polymers, in general.[8] By extrapolating the double log plot to t=1, $\log(Z)$ can be determined, however, it is convenient to use the half-life, $t_{\frac{1}{2}}$ to define the reciprocal rate of crystallization, since it corresponds closely to the time to attain the maximum rate of crystallization in the isothermal curves, and

$$Z = \ln(2)/t_{\frac{1}{2}}^n \qquad (10)$$

Further analysis requires the isothermal crystallisation to be split into primary and secondary processes.

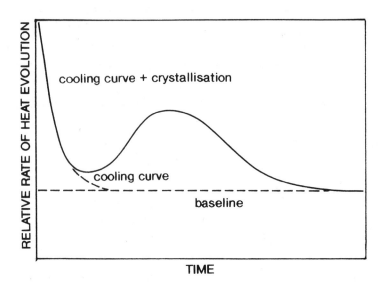

Fig.7   Composite cooling and crystallisation curve

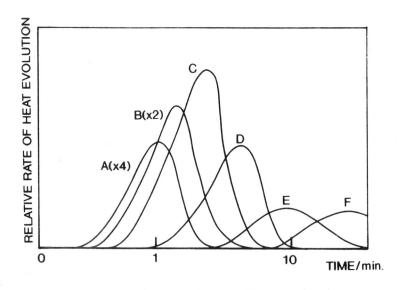

Fig.8a   Isothermal   crystallisation   curves   for
polyethylene.   A,147.7K; B,418.7K; C,419.7K; D,420.7K;
E,421.7K; F,422.7K

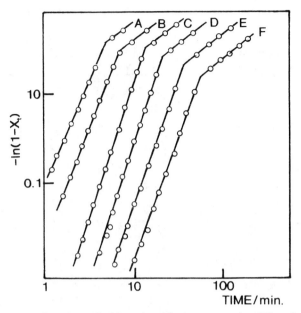

Fig.8b   Analysis of the isotherms as in Fig.8a

X_p is the final degree of crystallinity due to the primary processes, and is used as an adjustable parameter to give a constant n value over the primary process. An instantaneous value of n can be determined using,

$$n = t(dX_t/dt)[(1-X_t).ln(1-X_t)]^{-1} \qquad (11)$$

In Table 2, crystallisation rate data is listed for PEEK in the two crystallisation regimes. They exhibit very different temperature dependences in that in (a) the rate of crystallisation slows down as the melting point is approached, and the rate of crystallisation depends on the undercooling from the melting point, i.e. $(\Delta T)^{-1}$ where $\Delta T = T_m - T_c$. The rate determining effect is that of forming the critical size nuclei for growth to occur. In (b) the rate of crystallization decreases with decreasing temperature, and stops at the glass transition temperature. The rate determining step is that of segmental motion and movement of the chain to the crystal growth face. In this region the increased

Table 2 Crystallisation Rate Behaviour of PEEK.

Crystallisation Temperature/K	Avrami Rate Parameters			
	n	$t_{0.5}$ min.	z min.$^{-n}$	X_p
a) Low Temperature				
427.4	2.1	49.5	1.91×10^{-4}	0.80
428.4	2.1	40.2	3.12	0.95
429.5	2.2	22.9	9.66	0.85
430.5	2.3	18.4	18.40	0.95
432.5	2.3	9.30	9.18	0.90
b) High Temperature				
589.8	3.1	5.1	50.4	0.80
591.8	3.1	7.9	11.6	0.80
593.9	2.9	10.8	5.56	0.85
596.9	2.9	18.3	1.25	0.80
598.9	2.7	26.6	0.88	0.80
601.1	2.6	39.9	0.60	0.85

viscosity accounts for the slower rate and an Arrhenius equation can be used to fit the observed rate dependence. Once these temperature dependent rate equations have been empirically determined the overall crystallisation behaviour of a polymer can be simulated especially in thick sections when heat diffusion may be rate determining.

6 THE DEGREE OF CRYSTALLINITY.

Polymers are only partially crystalline but the degree of crystallinity, X_c, reflects their thermal history, i.e. crystallisation temperature, time, annealing temperature, etc. Physical and mechanical properties vary markedly with X_c and can limit their commercial and technological exploitation. Most approaches to the measurement of the degree of crystallinity adopt a two phase model, made up of crystalline and amorphous regions. This is undoubtedly an oversimplification, but nevertheless it is convenient and universally adopted. X_c can be determined by many experimental procedures but each adopts different definitions of crystallinity and measures different values for X_c.

DSC determines X_c from the enthalpies of fusion or crystallisation, ΔH_x, by measuring the area under the melting curves, i,e.

$$X_c = \Delta H_x / \Delta H^o_x \qquad (12)$$

where ΔH^o_x refers to the enthalpy of fusion or crystall- isation of the totally crystalline sample. It should be appreciated that both enthalpies are temperature depen- dent and both should be measured at the same standard temperature, usually T^o_m. This is seldom done, and ΔH^o_x is difficult to measure. It is difficult to determine the heat capacity temperature dependence of the cryst- alline solid between the crystallisation and melting temperatures, since annealing may be occurring. This makes the baseline and the integration of the fusion peaks uncertain. The method, however, is widely used to measure relative crystallinities for comparison of samples and is also used calibrate against samples whose degrees of crystallinity have been determined by other techniques, such as WAXRS, density, IR or NMR spectro- scopy. Care has also to be taken in interpreting the residual crystallinity obtained in quenched "amorphous" samples. Direct comparison cannot be made of the enthalpies of fusion and crystallisation, because of their temperature dependence,

$$\text{i.e. } -\Delta H_c = \Delta H_f - \Sigma^{T_c}_{T_f} \Delta C_p dT \qquad (13)$$

since they may differ by as much as 25-50% in value.

Direct evaluation of the enthalpy of the sample between temperatures T_c and T_f will, however, enable the residual crystallinity to be determined since by the First Law subsequent melting and recrystallisation are unimportant and it is only the difference between the enthalpies of the amorphous and quenched sample which is important, see Figure 9. In this way it was established that quenching thin films of PEEK in liquid nitrogen produced totally amorphous film, within experimental error.

7   MELTING   BEHAVIOUR   AND   THE   QUALITY   OF CRYSTALLINITY.

Low molecular weight ultra-pure materials have a single-valued   sharp   m.pt,   which   is   characteristic   and determined by equilibrium thermodynamics:

$$\text{i.e. } T_m = \Delta H_f / \Delta S_f \tag{14}$$

This is not the case in polymers. Firstly melting occurs over a wide temperature range, 10-50K, and the m.pt. is not   unique   in   that   it   varies   with   crystallization temperature. If a m.pt. is required then usually the last   trace   of   crystallinity   is   taken,   not   the   peak temperature and not the first onset of melting. Measured anyway,   the   observed   m.pt,   increases   with crystallization temperature, see Figure 10.

Using the fringe micelle model Flory[9] explained the dependence since on heating, smaller crystals melted and subsequently   recrystallised,   but   as   thicker   crystals with   an   even   higher   melting   point.   At   higher temperatures the rate of crystallization was slower and more   perfect,   larger   crystals   were   formed.   Rapid crystallization at lower temperatures produced less well formed,   imperfect   smaller   crystals.   Accordingly   to obtain a meaningful m.pt. the sample had to be heated slowly to allow the crystals to perfect and larger ones to develop. Typically a heating rate of $1°$ a day was adopted, and the equilibrium m.pt, $T°_m$, was defined as the   m.pt   of   the   perfectly   ordered   crystal   of   an infinitely large chain. The observed m.pt would always be less than the equilibrium m.pt since the molecules did not have infinite molar mass. Chain ends lower the m.pt. and Flory derived the following relationship,

$$1/T_m - 1/T°_m = [R - (1+b)]/\Delta H x_n \tag{15}$$

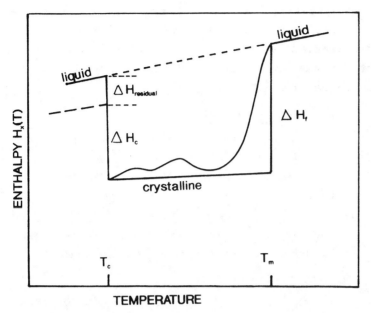

Fig.9 Schematic representation of the enthalpy changes
on melting and recrystallisation

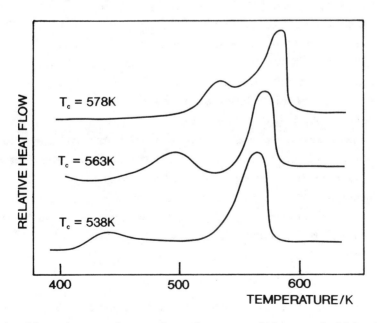

Fig.10 Change in m.pt. of PEEK with crystallisation
temperature

where $b = (1-(l_e-1)/x_n)$, l_e is the equilibrium length of the crystal, x_n is the number average degree of polymerisation, and ΔH is the heat of fusion per monomer unit. For the most probable distribution of molar mass this reduces to

$$1/T_m - 1/T^o_m = 2R/x_n\Delta H \qquad (16)$$

Plots of $1/T_m$ against x_n while linear gave a poor estimate of ΔH and there is little doubt that the experimental procedure underestimates the equilibrium values of the m.pt.

Using the nucleation theory of lamellar growth, Hoffmann and Weeks[10] have developed a relationship for the m.pt and the crystallization temperature, T_c,

$$T_m = T_m^o(1-2/\beta) + T_c/2\beta \qquad (17)$$

where β is the ratio of stem length, l, to equilibrium stem length, l_e and fold surface energy, σ, to equilibrium surface energy, σ_e.

$$\beta = \{1_e\sigma/1\sigma_e\} \qquad (18)$$

Plots of T_m against T_c were linear with slopes of $1/2\beta$, and when extrapolated cut the line $T_m = T_c$ at T^o_m. If no annealing occurred then $\beta=1.0$ and the slope was $1/2$, see Figure 11. Mandelkern et al.[11] have analysed this equation in depth and shown that it underestimates T^o_m unless the polymers have molar mass above 100,000, i.e. the end groups lower the m.pt.

Lamellar crystals of polyethylene are very similar in structure to the n-alkane crystals, and these have been used as model compounds to interpret the melting of polyethylene[12]. Differences arise in that the fold surface is covered with $-CH_3$ groups rather than chain segments. With the n-alkanes, C_nH_{2n+2}, the heat of fusion, ΔHn, is a linear function of n,

i.e. $$\Delta Hn = n\Delta H + \Delta H_e + n\Delta C_p dT \qquad (19)$$

subscript e refers to the terminal $-H$ units, ΔC_p is the heat capacity difference between the melt and the

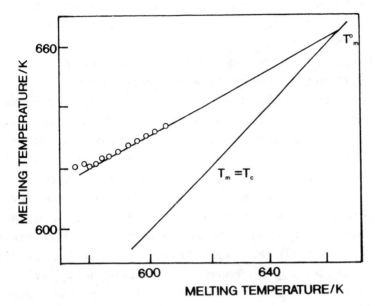

Fig.11   Dependence   of   m.pt.   on   crystallisation
temperature

crystal per $-CH_2-$ unit, and ΔT = T^o_m - T_m. The
corresponding term for entropy change, ΔS_n is

$$\Delta S_n = n\Delta S' + \Delta S_e + 2R\ln(n) + \Delta C_p dT/T \qquad (20)$$

the subscripts having the same association, and $\Delta S'$
refers to T_m. $2R\ln(n)$ is the entropy of mixing terminal
units with the repeat unit. Since at T_m,

$$\Delta G_n = \Delta H_n - T_m\Delta S_n = 0 \quad \text{and} \quad \Delta G = 0 \qquad (21)$$

$$\text{then} \quad T^o_m = \Delta H/\Delta S \qquad (22)$$

T^o_m refers to the infinite crystal with no end groups.
Combining these equations, gives

$$T_m = T_m^o(1-2RT_m\ln(n)/n\Delta H)- \sigma_e/n\Delta H - \Delta C_p\Delta T/n\Delta H \qquad (23)$$

in which σ_e = $-\Delta G_e$ = $T\Delta S_e$ - ΔH_e. The 3rd and 4th terms
are small in comparison to the 2nd, and setting $T_m/\Delta H$ =
$T^o_m/\Delta H^o$, the equation truncates to,

$$T_m = T^o_m(1-2RT^o_m\ln(n)/n\Delta H^o) \qquad (24)$$

Plots of m.pt against $\ln(n)/n$ for the n-alkanes and

polyethylene oligomers are linear with slopes $2RT^{o\,2}_{m}/\Delta H^{o}$
and intercept T^{o}_{m}, see Figure 12. The procedure shows
that the m.pt of a crystal is determined by its
thickness and since the crystal thickens with increasing
crystallization temperature the dependence of T_{m} on T_{c}
arises. The above analysis is for single component
hydrocarbons or oligomers of low D.P and polydispersity.
It has been applied to polymers and there is coincidence
between the observed m.pt, and crystal thickness for the
n-alkanes and polyethylene lamellae, see Figure 12.

The shape of the melting endotherms is considered
to reflect the crystal size distribution and equation 23
has been used to analyse the endotherms and determine
the quality of the crystallinity within the sample.
Annealing at higher temperatures increases the range of
crystal thickness and so increases the m.pt and improves
the quality of the crystallinity.

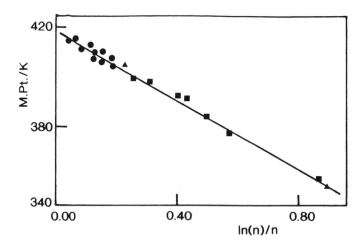

Fig.12   Dependence of the m.pt. of polyethylene on
degree of polymerisation, n. ● polyethylene;
▼ ethylene oligomers; ■ n-alkanes

8 CONCLUSIONS.

From the few limited examples listed above it is apparent that in the characterisation of thermoplastic materials thermal analytical techniques, particularly DSC, are extremely powerful and flexible techniques. To use them to their full capacity the heat capacity/ temperature or time data must be available for processing. Unfortunately this is not always the case with subsequent loss in effectiveness.

9 REFERENCES

1. L.Aras and M.J.Richardson, <u>Polymer</u>, 1989, <u>30</u>, 2256.
2. J.N.Hay, <u>Progress in Colloid & Polymer Sci.</u>, 1991, <u>87</u>, in press.
3. G.Williams and D.C.Watta, <u>Trans. Farad. Soc.</u>, 1970, <u>66</u>, 80.
4. M.L.Williams, R.F.Landel and J.D.Ferry, <u>J.A.C.S.</u>, 1955, <u>77</u>, 3701
5. A.A.Goodwin and J.N.Hay, <u>Polymer Comm.</u>, 1990, <u>31</u>, 388.
6. D.J.Kewmish and J.N.Hay, <u>Polymer</u>, 1985, <u>26</u>, 905.
7. M.Avrami, <u>J. Chem Phys.</u>, 1939, <u>7</u>, 1103; ibid, 1940, <u>8</u>, 212 and 177.
8. A.Booth and J.N.Hay, <u>Polymer</u>, 1961, <u>4</u>, 61.
9. P.J.Flory, <u>J. Chem Phys.</u>, 1942, <u>10</u>, 51.
10. J.D.Hoffman and J.J.Weeks, <u>J. Chem Phys.</u>, 1965, 42, 4301.
11. M.Gopalon and L.Mandelkern, <u>J. Phys. Chem.</u>, 1967, <u>71</u>, 3883.

Pharmaceutical Applications of Thermal Analysis

M. J. Hardy

SMITHKLINE BEECHAM PHARMACEUTICALS, R & D TECHNOLOGIES, THE PINNACLES, COLDHARBOUR ROAD, HARLOW, ESSEX CM19 5AD, UK

1 INTRODUCTION

The development of a new ethical pharmaceutical, from the initial synthesis to a licence application, can take between eight and twelve years. Thermal Analysis has many potential applications during this period, ranging from support for the initial fundamental chemical or biological research, through preclinical development to the final stages of licence application.

Rather than present a list of thermal analysis techniques which have use in pharmaceutical development, I propose to follow the path of a new drug from initial discovery to product launch, showing how Thermal Analysis can help in the full characterisation that is required. I will illustrate this with examples from various SmithKline Beecham laboratories, and the literature. The list is by no means exhaustive, but does cover the major areas in use today.

2 RESEARCH BIOLOGY

In fundamental pharmaceutical research, the initial requirement is for a biological model of the target disease state. This is the starting point from which all the chemistry must flow. It is therefore the logical place to start in the consideration of pharmaceutical applications of Thermal Methods.

All drugs, in order to have any activity, must interact with the target biological system. If this interaction can be measured qualitatively and quantitatively, then useful information can be generated possibly on the relevant structure-activity relationship.

Aqueous lipid dispersions have been used by a number of workers as models for biological membranes. An example of this type of work was recently published by Reid et al. in Biochimica et Physica Acta [1]. This paper described the use of, amongst other techniques, DSC for the characterisation of the interaction of model H^+/K^+ ATP-ase inhibitors with a dimyristoylphosphatidyl- choline (DMPC) model membrane. Figure 1 shows the effect that various inhibitors have on the transition onset temperature and the shape of the endotherm of DPMC .

When these results were considered with the FT-IR and ^{2}H-NMR data, it showed that the inhibitors readily partition into phospholipid bilayers, whilst not having any effect on the phospholipid bilayer structure itself. The main effect was a reduction in the co-operativity of the gel to liquid crystalline phase transition.

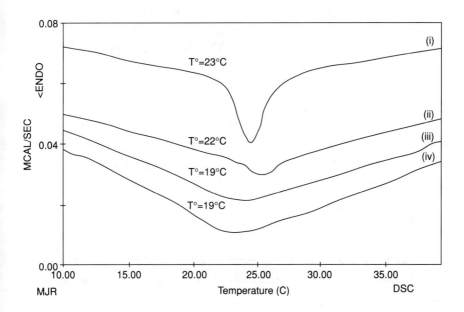

Figure 1 - DSC Curves of DMPC (i) and DMPC mixed with various inhibitors in the ratio 8:1

3 RESEARCH CHEMISTRY

The aim of the Research Chemist is to synthesise the target molecules suggested by the structure - activity relationships of the accepted model of the diseased state. The initial priority is for the confirmation of the structure of the target compound. Thermal methods cannot help in answering this question, but they can help in a number of related areas.

Structural Characterisation and Purity

As already mentioned, the first analytical task required by the research chemist is for confirmation of structure. This will usually be by the use of techniques such as ^{1}H and ^{13}C NMR spectroscopy, Mass spectrometry and elemental microanalysis. Together these techniques give some idea of purity, but the classical assessment of purity is usually by the determination of the melting range (it is not until the development phase that specific chromatographic assay methods are required). DSC or DTA are ideal for this routine task, although DSC is preferable, as this also gives the ΔH_f of the melt. This can be most useful for future comparison with further batches.

Stoichiometric Measurements

Thermogravimetry is very useful at this early stage, where hydration or solvation is suggested from the elemental microanalysis. The presence of a volatile can be trivially confirmed with analysis of the evolved gas either using mass spectrometry, or a liquid nitrogen trap, followed by solvent or water analysis.

4 CHEMICAL DEVELOPMENT

Once a compound is accepted into Development, the initial priority is to confirm that the compound is safe to give to healthy volunteers. This requires that a significant quantity of material is available for the toxicity testing. The synthetic route will need to be modified to produce the substantial quantities of compound that will be required. Analytical work focuses on the development of analytical methods for the clearance of material, the assignment of a suitable standard(s), assessment of stability and the generation of the batch analysis data that is required by the regulatory authorities.

Thermal analysis has an important role in this phase. The DSC purity method is of use for the assignment of standards and method validation, but not usually as a routine method. The potential of the compound to exist in various crystal forms must be investigated; thermal methods have important applications in characterising polymorphs and other solid state forms (amorphous forms, hydrates and solvates). Both DSC and TG can also have use in the assessment of stability. The relevant thermal analytical techniques will now be discussed in more detail.

Purity

Melting point determinations have long been used as a method of purity assessment of organic compounds. These compounds could then be classified as "pure" if the melting point was sharp, or "impure" if there was a broad melting range [2]. The DSC method which effectively quantifies this procedure, is based on the well known Van't Hoff equation:

$$T_s = T_0 - \frac{R\ T_0^2\ X_2}{\Delta H_f} \times \frac{1}{F}$$

where:-

T_s = Sample temperature (K)

T_0 = Theoretical melting point of pure compound (K)

R = Gas constant (1.987 cal mol^{-1} K^{-1})

X_2 = Total mole fraction impurity

ΔH_f= Heat of fusion of pure compound (cal mol^{-1})

F = Fraction of sample melted at T_s

The equation will only be valid under the following circumstances [3]:

1. The compound does not decompose at or near its melting point.
2. The impurities form an eutectic with the main component, i.e. the impurities are soluble in the liquid phase of the main component.
3. The impurities are not soluble in the main component in the solid phase (i.e. no solid solutions are formed).
4. The system is at constant pressure.
5. The heat of fusion is independent of temperature.

6.   Ln $X_1 = -X_2$. This effectively limits the determination to samples of >95 mol % purity.

7. $T_sT_0 = T_0^2$. This will be true at all but cryoscopic temperatures.

8.   There are no other thermal events in the vicinity of the melting region, i.e. no volatile losses or polymorphic transitions.

The necessary data can be generated by melting an encapsulated sample of 1 to 3 mg by a slow (0.5 to 2.0°C min^{-1}) temperature program. The melting endotherm is recorded, then segmented, allowing for the thermal resistance of the DSC system by the use of standard pure materials such as indium. The area of the segment is then related to a temperature (T_s) and the fraction melted is calculated by ratio with the total area of the melting endotherm. The temperature T_s is then plotted against the reciprocal of the fractional area. Theoretically this plot should be linear, but in practice this is rarely true. Various reasons have been suggested for this deviation, such as:

1.   Thermal lag between the sample and sample holder.

2.   Sensitivity limitations of the DSC allowing undetected premelting.

3.   Formation of solid solutions.

4.   Regions of non-crystalline or amorphous material within the sample.

However, whatever the cause a correction must be made to obtain the necessary linearisation. This is usually a trial and error addition to both the fractional and total areas (often quoted as a percentage of the total area). The corrected total area is then utilised to calculate the heat of fusion for the compound. The heat of fusion and the slope of the corrected T_s vs 1/F plot give the mole purity. The DSC purity data can be used in two main ways. Firstly for the assignment of reference standards, and secondly to help with the accuracy validation of other purity methods.

Assignment of Standards.   Early in the development process, we will be required to routinely monitor batches of both bulk drug and finished products (ie. a tablet, capsule etc.) for purity. The assay method is usually chromatographic and therefore relative - hence the need for an analytical standard for comparison. This material will usually be as pure as possible and assigned by subtracting the total percentage of all unknown and known purities from 100%. However alternative checks on these assignments are required before authorisation; this allows not only a critical judgement of the suitability of the standard but also of the chromatographic techniques used in the primary assessment.

DSC purity analysis can usefully be applied to standards, provided of course that the melting-curve is ideal, and other limitations of the Van't Hoff equation are satisfied. Assigning mole percentages can present problems. For the high purity compound, although the impurities are unknown the mole percentage figures will usually not differ much from the wt/wt percentages. Results usually show a good comparison between the assigned value of the standard and the subsequent assay of another batch by HPLC.

Table 1 - Comparative data for the Assignment of three Reference Standards of a Development drug

	Assigned Purity (%w/w)	Volatile Imps (%w/w)	Other Imps (%w/w)	HPLC Assay (%w/w)	DSC Purity (mole%)	DSC Purity *Pin-hole (mole%)
Standard 1	99.3	0.1	0.6	-	98.4	99.8
Standard 2	99.4	0.1	0.4	99.8	99.0	100.0
Standard 3	99.4	0.1	0.4	99.4	98.5	99.7

*see text for explanation

Table 1 shows a set of data generated on three progressive standards of the same development candidate. The initial standard was assigned 99.3% w/w on the 100-(Total Detected Impurities)% basis. However, the DSC purity figure of 98.4 mole% was significantly different from this figure. Because the melting point of the compound was high (c.a. 290°C), the run was repeated with the lid of the sample holder pierced with a "pin-hole", to allow the escape of any volatilised impurities. This gave a more realistic result of 99.8 mole% when compared to the assigned figure. This was further evaluated when the second and third standards were prepared. Table 2 shows the effect of performing the DSC purity calculation using different portions of the melting endotherm to obtain the partial areas, and clearly shows that the compound decomposes after it has been annealed at temperatures above 200°C. However the rate is slow, so that the purity figure is satisfactory, unless the compound is held for a significant time above 200°C.

Validation of Accuracy. The DSC method can also be used to validate an HPLC method for accuracy, as it is most encouraging to have two methods based on different principles which give consistent results. Table 3 shows the analysis of ten different batches of drug by both DSC and HPLC. The assigned purity given to each batch is obtained by subtracting all detected impurities from 100%, which allows comparison of the two methods.

Table 2 - The effect of the choice of the part of the endotherm used in the Purity Calculation.

Annealing Conditions for Standard 3 with pin-hole*	Purity with 6% to 59% of Endotherm used (mole%)	Purity with 6% to 25% of Endotherm used (mole%)	Purity with 25% to 40% of Endotherm used (mole%)	Purity with 40% to 59% of Endotherm used (mole%)
None	99.7	-	-	-
15min @150°C	99.8	99.8	99.8	99.9
10min @200°C	99.2	-	-	-
10min @240°C	99.1	99.2	98.7	97.7

* See text for explanation.

The main impurities in these batches were water and propan-2-ol. These were not detected by DSC because the solvent impurities had vapourised before the melt (at ca. 116°C), which resulted in these impurities not being soluble in the liquid phase of the drug. The main organic impurities do have very similar molecular weights to the main component (365.5) and therefore we felt justified, in this case, of assigning the w/w percentage as equivalent to the mole percentage that would be detected by DSC.

The results of the analyses are given with the variation from the assigned purity in parenthesis. These show good agreement between methods, and therefore the DSC results validate the HPLC method for accuracy (and vice versa, of course!).

Polymorphism and Solid State Forms

Polymorphism is the name given to the ability of any compound to exist in more than one crystalline species. This phenomenon is of great importance, as each polymorph may have different physical properties, such as density, hardness, solubility and melting point and often the polymorphs have differing stabilities. However Pharmaceutically, we are interested in all solid state forms, which will also include amorphous forms (i.e. glasses) and solvates and hydrates (i.e. where either solvent or water is incorporated into the crystal structure). The reason that this is so important, is that at any one given temperature and pressure, one form will be thermodynamically more stable. Therefore less stable forms may convert

Table 3 - Comparison of DSC and HPLC data for a "low" purity compound (ca. 96.0%w/w)

Batch	Assigned purity (%w/w)	Purity by DSC (mole%)		Purity by HPLC (%w/w)	
5001	96.0	96.9	(+0.9)	95.8	(-0.2)
5002	96.6	95.0	(-1.6)	95.7	(-0.9)
5003	95.0	93.7	(-1.3)	94.5	(-0.5)
5004	96.5	96.5	(0)	96.2	(-0.3)
5005	96.8	98.1	(+1.3)	97.6	(+0.8)
5006	95.7	96.0	(+0.3)	96.0	(+0.3)
5007	96.1	96.4	(+0.3)	95.7	(-0.4)
5008	96.0	96.5	(+0.5)	96.4	(+0.4)
5009	96.1	97.0	(+0.9)	96.1	(0)
5010	95.4	95.4	(0)	96.0	(+0.6)
-	-	±1.3 from assigned purity		±0.7 from assigned purity	

immediately or under conditions of relatively mild stress. It is therefore essential that any propensity to form multiple solid state forms must be characterised as soon as possible. Examples of problems are numerous, but include:

1. The precipitation of a hydrate from a stored aqueous solution of a drug - the hydrate was soluble at about 2 mg/ml, whilst the solution was 10 mg/ml of the anhydrous material (solubility of >50 mg/ml).
2. The conversion of one polymorph to another in tablets which were being stored under standard conditions - due to the less stable form being tabletted initially.
3. Gradual conversion of an apparently stable anhydrous crystalline form under high relative humidity conditions to a hydrate. This hydrate partially dehydrated in low humidity conditions.

However, our aim is to characterise as many forms as possible. Initially as the first batches are produced, they are characterised using DSC and Nujol mull IR. If differences are seen, the search stage is initiated. This entails a recrystallisation programme, using all synthetically and pharmaceutically relevant solvents, attempting to vary any relevant parameters (e.g. speed of crystallisation, temperature of crystallisation etc.). The products are then characterised by appropriate analytical techniques selected from DSC, IR, solid state ^{13}C NMR, Laser Raman, powder X-ray diffraction, microscopy and thermomicroscopy. If differences are reproducible, larger quantities are produced for further pharmaceutical characterisation and often a single crystal X-ray will be determined, so that the differences in structure can be understood.

An example of where DSC can be of use is shown in figure 2. Form II is the stable form which melts at about 180°C, with no other transitions being observed. However as Form I is heated, a number of events are seen. Melting begins in the range of 115°C to 120°C, and is followed by a recrystallisation exotherm between 120°C and 130°C. Melting then reoccurs at about 180°C. On cooling and reheating, only the Form II melting endotherm is seen. This type of behaviour is often observed when one form is much less thermodynamically stable than the other (melting points 115°C vs 180°C) but there is a significant kinetic barrier to any solid state transitions. However, as soon as the isotropic liquid is produced as Form I melts, Form II spontaneously recrystallises from the melt, which then melts at its melting point of 180°C.

Thermogravimetry is often of use when hydrates (or solvates) are being studied, but often a number of thermal techniques need to be utilised to fully characterise complex systems [5]. Care must also be exercised in the selection of experimental conditions, e.g. pan type. Figure 3 shows such a system, with the TG curve (1) followed by three DSC curves in sealed (2), crimped (3) and open (4) pan configurations. The figure shows that the pan type used in the DSC experiments alters the onset of the dehydration endotherm(s), but this also masks other melting and recrystallisation transitions. Observations using thermomicroscopy are tabulated in table 4.

Table 4 - Thermomicroscopy observations for a development drug.

Temperature (°C)	Observation
112-130	Sample melts
140-170	Sample recrystallises
220	Sample melts with decomposition
240-260	Sample undergoes further decomposition

The effects seen in the DSC experiments are probably due to increased internal pan pressures, but it is interesting to note that the open pan run fails to detect the initial melting, possibly as it is a second order transition. Also the open pan shows that the dehydration is energetically complete by 80°C, whilst the actual

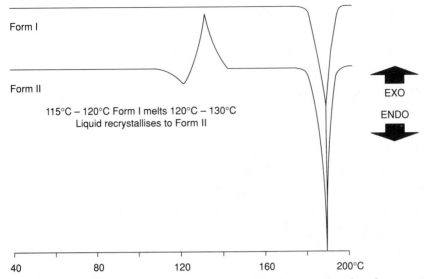

Figure 2 - DSC curves for two polymorphic forms of a development drug

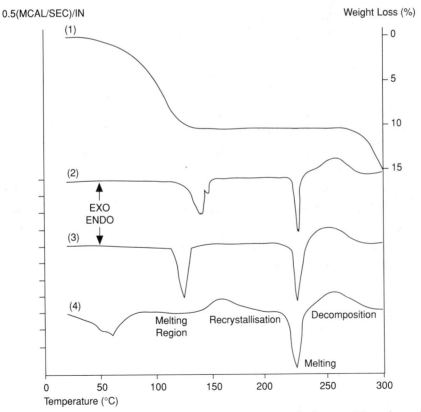

Figure 3 - The TG curve (i) and DSC curves in sealed pans (ii), crimped pans (iii) and open pans (iv) of a drug candidate

weight loss continues to 120°C, presumably due to the anhydrate being hygroscopic and so delaying the actual water loss.

These examples show the value of thermal techniques in the assessment of solid state forms. However it is more usual for the full range of techniques mentioned above to be used in an attempt to characterise the system as completely as possible.

Stability

Although real time stability data is always required, estimates of both the bulk drug and the formulated product stability are always generated. These are conventional storage tests utilising a variety of temperatures and relative humidities. The data is interpreted using the Arrhenius equation:

$$K = Ze^{-E/RT}$$

where:

K is the specific rate constant at temperature T

Z is the Arrhenius Frequency Factor

E is the Activation Energy

R is the Gas constant

Large quantities are stored and periodically assayed (main peak + impurities) by such technologies as HPLC, TLC, titration etc. and an estimation is made of the $t_{0.05}$ (time to 95% of original concentration).

The variable heating rate method of Ozawa is the most commonly applied method used for DSC stability prediction. The compound is initially run until decomposition is observed. If this decomposition occurs at or above the melting point, and/or is endothermic, the compound is considered to be stable (probably will have a shelf life of at least two years). If the decomposition is below the melting point and is also exothermic, a series of DSC runs are initiated with a range of heating rates. From the plot of $\ln \beta$ vs $(1/T)$, the activation energy (E) and the

Arrhenius frequency factor (Z) can be derived, as

$$E = \frac{- R \, d[\ln(\beta)]}{d \, [(1/T)]}$$

and

$$Z = \frac{\beta E e^{E/RT}}{RT^2}$$

where β is the heating rate and T is the DSC peak temperature.

The rate constants at various temperatures can then be simply calculated. The half lives may be tested by isothermal ageing at the required temperature. The area under the exotherm in the 'aged' sample is compared to that of an 'unaged' sample. Any large deviation from the predicted 50% can be taken as evidence that the method is not applicable in that particular case. Also a non-linear $\log \beta$ vs $(1/T)$ plot would indicate a similar non-applicability, (probably due to competitive reactions occurring at the elevated temperatures).

It is not recommended that this data is used for the calculation of any other kinetic data, as the nature of thermal analysis is dynamic, so rendering the assumption that the system is in equilibrium as most doubtful. If such data as " safe storage temperature " is required, the use of a true calorimetric method such as ARC (Accelerated Rate Calorimetry) is highly recommended.

5 FORMULATION DEVELOPMENT

Drug-Excipient Compatibility

The Pharmaceutical Handbook [6] defines incompatibility in a medicine as "an interaction between two or more components to produce changes in the chemical, physical, microbiological or therapeutic properties of the preparation". Chemical incompatibility will usually be due to redox, acid-base, hydrolysis or combination reactions; physical incompatibilities are changes in solubility, adsorption of a drug onto an excipient or formation of an eutectic; microbiological incompatibility is usually associated with the reduction in effectiveness of antimicrobial preservatives due to interaction with drug or excipient; and finally, therapeutic incompatibility is interaction of the drug with other drugs or foods taken concurrently, which effect changes in the therapeutic or toxic effect of the drug [3].

Once a drug is formulated into a dosage form, storage tests at various temperatures are inititiated. These tests can last up to five years and provide the final data on, amongst other things, the compatibility of the drug with the chosen excipients. This compatibility is assessed by the comparison of "Main Peak Assay" results and "impurity profiles" of the active ingredient with time, using techniques such as HPLC, TLC, etc. This work is time consuming and expensive so it is most desirable to predict any possible incompatibility at an early stage. The DTA or DSC methods will record chemical reactions which either adsorb or emit heat, changes of state, eutectic formation, etc., as a function of temperature or time. DSC curves of the pure drug substance and pure excipient are recorded. A simple additive superimposition of these curves is then compared to a DSC curve obtained from a well mixed 1:1 mixture of the drug and excipient.

If there are no differences between the theoretical and experimental mixture curves, thermal analysis will suggest that there is no interaction between drug and excipient, ie. there is no physico-chemical incompatibility.

However, problems of interpretation arise when there are differences in these curves. The drug and excipient may still be compatible, as the DSC technique necessarily requires elevated temperatures and these temperatures might induce reactions that do not occur at normal storage temperatures. (N.B. The conventional storage tests at elevated temperatures, which use Arrhenius plots give similar problems). Also, as the active/excipient ratio is not usually 1:1 a series of runs which more closely match the chosen ratio might be necessary. Interaction will usually indicate either chemical reaction, adsorption or eutectic formation. This interaction might be advantageous (eg. as a more desirable form of drug delivery system) or is an example of physico-chemical incompatibility.

DSC can be used to screen excipient candidates. If any interaction is observed, that excipient can be avoided in favour of one that shows no interaction. Problems have also been encountered with false positives and negatives, so great care is necessary in the interpretation of results, otherwise useful excipients could be discarded erroneously. Also, interaction might indicate either complex or eutectic formation. This could be investigated (particularly if the drug has a low solubility) to find an alternative drug delivery system.

Phase Diagrams

The construction of phase diagrams is of great importance to the pharmacist in his study of polymorphism and dosage forms. The DSC can be of considerable help. For dosage form study, binary mixture curves can be set up. The various w/w or molar percentage mixtures can be prepared, and then DSC curves for each mix generated. The temperature of any relevant melting endotherms is plotted against the percentage drug content in the mixture. Areas of interest can be 'expanded' by running further mixes. Figure 4 shows an example of an attempt to formulate a drug candidate in a polyethylene glycol. The idea was to obtain a suppository formulation which would be readily absorbed. The phase diagram clearly shows that the drug - excipient mixture only melted as a single phase at 45°C with a maximum drug content of 10%. At higher drug incorporation a biphasic melt occurred. This study clearly showed that using this excipient the formulation would not work, as the 10% incorporation would not melt at 37°C.

Freeze Drying

Biological agents and antibiotics are often prescribed as reconstitutable formulations, which often require that free-drying is necessary during the final product preparation. The freeze-drying process depends for success on the conditions of the freeze-drying cycle as these will affect the physical appearance, chemical stability and ease of reconstitution of the finished product. These problems can be avoided by identification of eutectics, reversible and irreversible solid-solid transitions and super-cooling characteristics of the sample solution, so allowing a suitable freeze-drying cycle to be designed.

DSC is an ideal technique for identification of these transitions. By the judicious use of heating and cooling cycles varying the heating or cooling rate, the various phase transitions can be detected. Suitable isothermal annealing phases can be included, if solid-solid transitions are observed and results compared to trial freeze-drying runs or runs on a freezing analyser.

One example was for a 20% aqueous solution of a penicillin. Figure 5 a) shows a DSC curve for frozen water. The sample was pre-cooled with liquid nitrogen to -100°C, then heated at 5°C min. Only two solid-solid transitions can be observed. Figure 5 b) shows an identical run for the 20% solution. It can be seen that there are two very minor extra transitions at -51°C and -43°C and one 'major' transition at -35°C.

Initial freeze-drying runs showed that if the initial shelf temperature was greater than -37°C then the plug did not dry properly and melt-back occurred. However, an initial shelf temperature below -37°C produced a near perfect plug. The transition at -35°C has since been identified by a freezing analyser run, to be a eutectic. The fairly simple DSC run highlighted the possible problems.

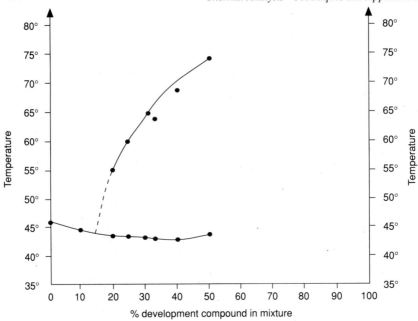

Figure 4 - Phase diagram for drug - Polyethylene glycol mixtures

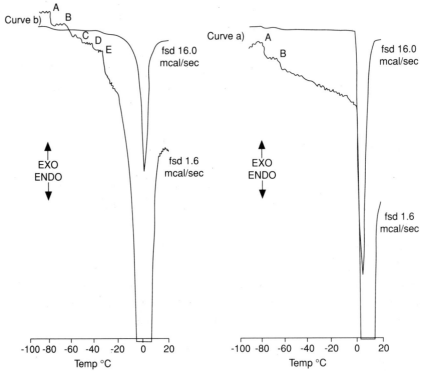

Figure 5 - DSC curves of a) Water and b) 20% w/v aqueous penicillin solution

However some caution must be observed as very small peaks are being considered. Solutions should therefore be of a high a concentration as is practicable and sample size fairly large (about 10mg) so that the signal to noise ratio is optimised.

Other studies have been concerned with a modified protein. An irreversible transition was detected from the frozen solutions. The DSC cell was cooled to about -75°C without a sample present. A sample was then placed in the cell so that the maximum cooling rate would be achieved ('flash' cooling). The sample was then heated at 10°C min^{-1}. Figure 6 a) shows that a fairly large transition was observed at about -20°C. If the sample was then allowed to cool back to -70°C (labelled 'slow' cooling) and the DSC curve regenerated (figure 6 b)), this transition does not appear. The initial freeze-drying of the protein did prove to be rather unsuccessful with long drying times and poor plug appearance. However the transition which occurs in the 'flash' cooled sample can be removed by a suitable annealing temperature. Figure 6 c) shows a DSC curve of a 'flash' cooled sample which has been heated to -15°C and held at this temperature for 30 minutes. The sample was then slowly cooled to -70°C and then the DSC curve generated. This annealing period has removed the transition which occurred at -20°C. Once this 30 minute annealing period was included in the freeze-drying cycle, a near ideal plug was obtained and this cycle was adopted for all further batches, with complete success.

To conclude; our initial work on frozen solutions has indicated that a lot of basic physical information is easily obtainable from the various DSC curves. The temperatures at which eutectics melt or solid-solid phase transitions occur can be readily translated to initial shelf temperatures and annealing temperatures respectively, which removes the more random approach that is sometimes applied.

6 CONCLUSION

Whilst thermal analysis cannot claim to be as widely used as spectroscopic or chromatographic techniques, it is of great use in general screening and trouble shooting at all stages of drug development, playing a critical role in pharmaceutical analysis.

ACKNOWLEDGEMENTS

I should like to thank Dr. Trevor Lever and Mr. Mike Raw, for running some of the DSC curves used to generate data in this review.

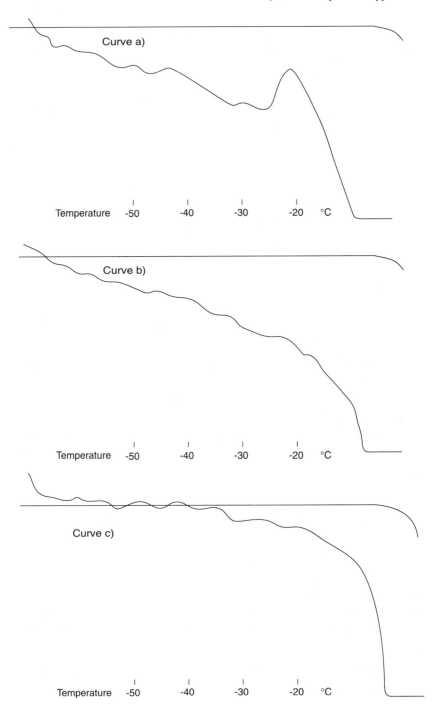

Figure 6 - DSC Curves of frozen protein solutions: a) Flash cooled,
b) Slow cooled and c) Flash cooled, then annealed at -16°C for
30 minutes

REFERENCES

1. D.R.Reid, L.K.MacLachlan, R.C.Mitchell, M.J.Graham, M.J.Raw & P.A.Smith, Biochim. Biophys. Acta, 1990, 1029, 24-32.
2. L.Kofler & A.Kofler, "Thermomikomethoden zur Kennzeidung Organischer Stoffe und Stoffgemische", Verlag Chemie, Weinheim, (1954)
3. M.J.Hardy, Thermal Analysis, Wiley Heyden, 1982, 876-892.
4. J.Haleblian & W.McCrone, J. Pharm. Sci., 1969, 58, 911-929.
5. D.E.Brown & M.J.Hardy, Thermochimica Acta 1985, 90 149-156.
6. Pharmaceutical Handbook 19th edition (1980) p 28.

Applications in Metallurgy and Materials Science

F. R. Sale and A. P. Taylor

MANCHESTER MATERIALS SCIENCE CENTRE, UNIVERSITY OF MANCHESTER AND
UMIST, GROSVENOR STREET, MANCHESTER M1 7HS, UK

1. INTRODUCTION

The range of TA applications that could be included under this title is enormous
and ranges from the industrial processing of minerals in extractive metallurgy,
through studies of alloy stability and ceramic production to the processing and
properties of polymeric materials. However, as earlier, separate papers in this
volume are devoted to minerals and polymers these aspects of materials science
will not be considered here. The main areas of application of TA that have been
selected for this paper are concerned with the use of DSC in the study of phase
transitions and heat treatment procedures in commercial alloys, the use of TG and
dilatometry in the evaluation of techniques for the production of electronic
ceramics and finally the use of STA in the study of intumescent coatings.

These three areas of applications have been selected to demonstrate a little
of the vast range of applications of thermal analysis techniques to materials
science. However, each area is a large topic in its own right and further subject
restriction is necessary within the space available. Accordingly, the alloy
development will be restricted to aluminium alloys, in particular drawing attention
to the use of DSC in the study of precipitation strengthening mechanisms which
are so vital to the current range of commercial alloys used in the aerospace
industry. However, it is appropriate to indicate that the relatively recent increase
in operating temperature of commercial heat flux DSC now allows the quantitative
study of similar reactions in high temperature metallic systems such as steels, cast
irons and refractory metal alloys.

One important growth area in the production of high grade electronic ceramics is the use of gel processing techniques which transfer the homogeneity of aqueous solutions to the solid state. This is achieved by gel formation and gel drying to give a precursor which is subsequently decomposed/oxidised to yield fine ceramic powder. The thermal stability and mechanism of decomposition/oxidation of the precursor are of vital importance in achieving control of homogeneity and particle size. TG will be demonstrated to be a convenient method for such studies as both the temperatures of instability and the mechanism of decomposition of a precursor can be determined. Additionally, it will be shown that dilatometry may be used to study the sintering of the powders and hence give data which allow the conditions of precursor decomposition to be related to the final stage of processing of the ceramic powder. The ceramics discussed here will be limited to high T_c superconductors.

The final part of this article will demonstrate how STA (TG/DTA) can be used to study the reactions between the active ingredients in intumescent coatings. These coatings, which expand to 80 or 90 times their original dimensions on heating, are used for the protection of structural steelwork. The critical, temperature sensitive reactions which cause the intumescence to occur need to be optimised in relation to char formation and ultimately to the insulation given to the steelwork. Examples will be discussed of the thermal behaviours of acid sources, carbonifics and spumifics used in such coatings. The results of thermal studies of mixtures of these compounds will then be used to discuss the mechanisms of reaction involved in the behaviour of the intumescent coatings under fire situations.

2. PRECIPITATION IN ALUMINIUM ALLOYS

In the use of DSC to study precipitation reactions in aluminium alloys the first stage of interpretation of the experimental data is to separate out thermal events which are caused by the preceding heat treatments, and hence are related to the microstructural development of the alloys, from those which are associated with transitions which occur as the alloys are heated in the DSC. Both may offer vital information concerning the use of the alloys, however, it is clear that they must be distinguished so that the correct heat treatment procedures may be realised.

A typical precipitation sequence observed on ageing an aluminium alloy is given in Figure 1.

(1)   Supersaturated solid solution

↓

(2)   Zones (G.P.)

↓

(3)   Intermediate precipitate

↓

(4)   Equilibrium precipitate

Figure 1:   Generalised precipitation sequence for an aluminium alloy.

In real systems, however, the situation is generally more complicated than shown in Figure 1. For an aluminium 4 wt% copper alloy of the duralumin type the sequence involves 2 types of zones before the first intermediate precipitate is found. In all systems the essential first stage of such precipitation sequences is the formation of GP (Guiner-Preston) zones (1) which may be described as clustering of the solute atoms on certain crystallographic planes of the solvent matrix. Such zones may lead to the production of a coherent precipitate prior to incoherent intermediate and equilibrium precipitates. The temperature regions of stability of the various zones are of importance to physical metallurgists who may represent the critical temperatures of non-equilibrium and equilibrium phases on a phase diagram. Such data are determined experimentally by classical hardening curves, resistivity measurements or microscopy. However, these techniques are extremely time consuming and DSC in conjunction with TEM as a complementary technique may be used to advantage to study advanced and complicated alloys (2). For example the commercial development of the Al-Li-Cu-Mg based alloys has prompted investigations of the basic phase equilibria and transformations which can occur within this system and its component binaries and ternaries (3). The research has involved microstructural examination with DSC providing important additional information. Specifically a second level of metastability in the Al-Li system, within the $\alpha + \delta'$ phase field, has been proposed and interpreted in terms of GP zone formation at low temperatures (4,

5). DSC studies have concentrated on the binary system, on two of the component ternary systems and on the role of zirconium additions to the binary system and to various quaternary alloys (2).

Heat capacity measurements on the binary Al-2.8 wt% Li alloy in the as-quenched (AQ), naturally aged (NA) and artificially aged (AA) conditions are shown in Figure 2(a) where the dotted line represents the specific heat of the alloy estimated using the Neumann-Kopp prediction. In the case of age-hardening alloys formation of zones and precipitate phases is normally an exothermic process whilst their dissolution is an endothermic reaction. In the AQ condition the specific heat moves in an exothermic direction during the earliest stages of heating (A) as GP zones form. This behaviour is terminated by the endothermic reaction at 130°C (B) which is caused by the dissolution of GP zones of the Al-Li system or retrogression of fine δ' which forms rapidly at room temperature or during the quenching process. The subsequent exotherm at 170°C (C) is similar to those observed in previous work (4, 5) and is associated with δ' formation. The endotherm "peaking" at 300°C (D_2) is again consistent with previous thermal analysis studies on binary Al-Li alloys and is attributed to δ' dissolution.

TEM studies of the as-quenched material revealed the presence of superlattice reflections prior to any heating in the DSC; however no δ' particles could be clearly imaged in dark field at this stage. After being heated to 250°C in the DSC and quenched, samples contained a homogeneous distribution of δ' particles ($\sim$ 10nm in diameter). Similar TEM studies after heating to 300°C revealed fewer δ' particles 35nm in diameter (with very fine δ' ($<$4nm) distributed throughout the matrix, presumably having formed from supersaturated solution during the quench from 300°C); these observations are consistent with endotherm D_2 being due to δ' dissolution.

Microscopic observations of the naturally aged alloy revealed similar features to the as-quenched material but the δ' could now be clearly imaged ($\sim$ 5nm particles). The DSC traces showed some notable differences. The endotherm at 130°C became more pronounced and contained indications of two maxima (B_1 and B_2). In a similar fashion the δ' dissolution peak is made broader by the appearance of a shoulder on the curve at about 250°C (D_1). These curves are similar in form to those of Nozato and Nakai (4) who attributed B_1 to the dissolution of an additional GP zone phase and D_1 to the dissolution of δ' particles having short range order. Artificial ageing prior to thermal analysis results in the disappearance of the dissolution reactions which occurred early in the continuous heating cycle (B_1 and B_2 < 150°C) whilst the endothermic peak is more prominent and extends down to ~ 220°C (i.e. between D_2 and D_1). The change in shape of the endotherm from that of the as-quenched material (D_2) on artificial ageing is associated with δ' dissolution.

The addition of 0.1% Zr to the binary alloy gives rise to substantial changes to the thermal analysis traces (Figure 2(b)). In the AQ condition the low temperature exotherms and endotherms (< 150°C) seen in the binary alloy are no longer evident, but the exothermic reaction (C) at 170°C is much more pronounced. The endothermic reaction (D_2) associated with δ' dissolution is more symmetrical with the peak moving below 300°C to about 270°C. Again, this δ' dissolution endotherm broadens with natural ageing and develops a plateau between D_1 and D_2 after artificial ageing extending down in temperature to ~ 220°C as observed in the binary alloy. Artificial ageing also substantially reduces the δ' formation exotherm (C) at 170°C. The microstructural effects of Zr are twofold. The solution treated microstructure is changed from the recrystallised equiaxed form of the binary alloy to an unrecrystallised structure containing subgrains and coherent $ZrAl_3$ (β') particles. The latter act as excellent heterogenous nucleation sites for δ' precipitation upon natural, and artificial ageing (3). Other than that there are no obvious differences in the δ' distribution compared to the binary alloy, either during heating up to 250°C in the DSC or during artificial ageing at 185°C.

The use of DSC to study Al alloys can be extended to study mechanical work and its effect upon precipitation (6). Figure 3 shows heat capacity data for a 7075 series Al alloy; (Al - 5.92 Zn - 2.60 Mg - 1.62 Cu - 0.1 Cr - 0.19 Si - 0.13 Fe in wt%), in the states of (a) as quenched, (b) quenched and deformed 15% by rolling, and (c) quenched and deformed 50% by rolling. The low-temperature exothermic peak at ~ 90°C which is present in the data for the as-quenched alloy is absent for the alloy in the deformed state. The sample deformed by 15% gave a small endothermic peak at ~ 130°C which was followed by an exothermic doublet in the temperature range 200-250°C. Similar heat capacity data were obtained for the alloy which had been deformed 50%, although the exothermic doublet moved to a lower temperature as a result of the extra deformation of the alloy.

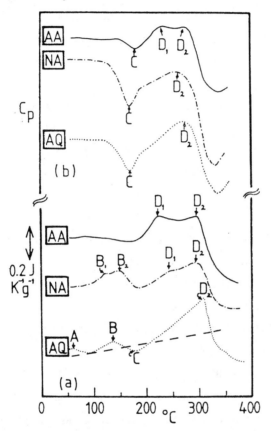

Figure 2(a): Specific heat capacity curves for Al-Li 2.8 wt% alloys

Figure 2(b): Specific heat capacity curves for Al-Li 2.6 wt% Zr 0.1 wt% alloys

AA Artificially Aged, 8h at 185°C; NA Naturally Aged, 3 months; AQ As Quenched from 540°C.

The disappearance of the low-temperature exotherm for the alloy in the deformed state indicates that the deformation of the as-quenched alloy has suppressed the formation of GP zones. It is known that vacancies play a major role in the formation and growth of GP zones (7, 8) and it is also acknowledged that climbing dislocations are effective sinks for vacancies. Thus it appears that the introduction of a high density of mobile dislocations in the supersaturated solid solution through the deformation process has markedly reduced the excess vacancy supersaturation and consequently decreased the rate of GP zone formation. The exothermic doublet seen for the deformed alloy is consistent with the heterogeneous nucleation of η phase. As the dislocation density is increased, by deformation, the size of the first peak on the doublet can be seen to increase. In the as-quenched alloy very little η phase is nucleated heterogeneously at 200-240°C because of the low dislocation density with the resultant very small 'shoulder' on the peak. The presence of a doublet for the deformed alloy may be associated with formation of η phase at two different types of heterogeneous nucleation sites, for example, dislocations and grain boundaries. The observation that the first peak of the doublet is displaced to lower temperatures as the amount of deformation is increased supports the concept that dislocations accelerate η precipitation.

Figure 4 shows heat capacity data for the alloy that has been aged 8 h at 120°C and for the alloy after the same ageing treatment but followed by 15% and 50% deformation. It is evident that the endotherm at ~ 200°C, which is associated with the dissolution of the η' phase, is reduced in size with increasing amounts of deformation. The most obvious explanation for this is that the amount of η' phase available for dissolution has also reduced, that is, reversion (or dissolution) occurs as a result of deformation. In addition to the reversion process the heating of the alloy during scanning in the DSC will anneal out dislocations which have been introduced into the material by rolling. This annealing-out process would be exothermic in nature and could negate part, or all, of the endothermic dissolution process if the temperatures of the two processes coincided. However, it may be expected from previous DSC studies of the recrystallisation behaviour of aluminium sheet (9) that the annealing process would occur at higher temperatures than that associated with the η' phase dissolution peak, and so the former explanation seems more likely.

Many more examples could be given showing the use of DSC in the study of metallurgical reactions. However, in the context of the present paper it is hoped that the preceding examples give sufficient evidence of the applicability and usefulness of thermal analysis in conjunction with TEM so that the reader may consider DSC as an essential tool in metallurgical studies.

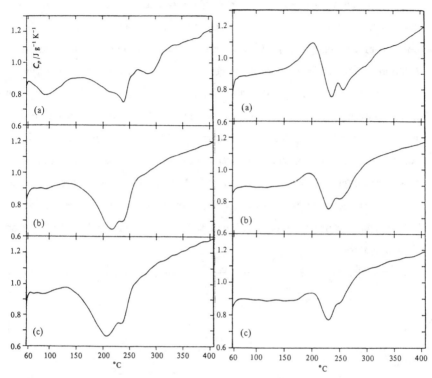

<u>Figure 3</u>:   <u>Figure 4</u>:

<u>Figure 3</u>: $c_p(t)$ curves for 7075 Al alloy; (a) as quenched and (b) quenched
            and deformed 15%; (c) quenched and deformed 50%.

<u>Figure 4</u>: $c_p(t)$ curves for 7075 Al alloy; (a) aged 8 h at 120°C; (b) aged
            8 h and deformed 15%; (c) aged 8 h and deformed 50%.

2. GEL PROCESSING OF HIGH T_c SUPERCONDUCTORS

The processing of a number of high technology ceramics has shown a need for
homogeneous powders of controlled particle size. Accordingly, a number of
specialised powder preparation techniques such as freeze-drying and gel synthesis
have been used to produce oxides with the required properties (10-16). The need
to produce homogeneous, pure superconducting oxide powders is clear from the
number of different reports on the properties of essentially the same material. In
addition, as impurities and grain boundary phases play important roles in limiting
the critical current-carrying capacities of these materials, pure homogeneous

powders are of great importance.

The citrate gel and the EDTA gel processes are used to obviate the experimental difficulties and chemical inhomogeneity associated with conventional ceramic processing (17-19). In the ideal situation, the homogeneity of a liquid solution is retained in an amorphous gel which, on subsequent pyrolysis, yields the complex oxide directly, providing that segregation and partial crystallisation can be prevented. Accordingly, it is vital to know the temperature and mechanism of decomposition of the precursors. TG and STA are ideal techniques for the study of the mechanisms of decomposition of the precursors particularly when used in conjunction with XRD and SEM to characterise the precursor and its products. As a further thermoanalytical technique dilatometry maybe used to measure the expansion and contraction of pellets of pressed powders as a function of temperature.

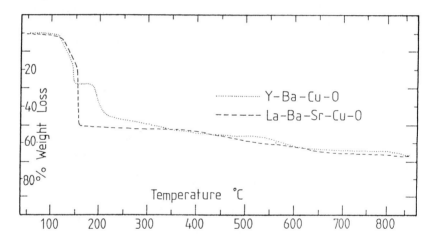

Figure 5:   TG data for pyrolysis of La- and Y-based precursors.

Figure (5) shows TG data obtained for the decomposition/oxidation of citrate gel precursors of $La_{1.7}Sr_{0.15}CuO_4$ and $YBa_2Cu_3O_{7-x}$ (20). It is evident that the two precursors decompose in different manners. The precursor of the La-based superconductor decomposes by a single-stage process once dehydration is complete at approximately 150°C. The precursor of the Y-based superconductor can be seen to decompose in two separate stages. For each precursor there is a slight weight loss on heating up to 450-550°C which is associated with the removal of residual carbon and in the case of the Y-based material with the loss of oxygen from the superconductor. Previous work by Courty et al (21) recognised the existence of two types of pyrolysis reaction for citrate-nitrate precursors. Type I was a single stage process for precursors which contained metals with a strong catalytic activity in oxidation processes. Such a pyrolysis obviously occurred for the La-based material. A second pyrolysis behaviour, Type II, was typified by a two-stage process in which an intermediate decomposition step occurs as a result of the formation of a metastable semi-decomposed precursor such as a mixed citrate salt. The step in the decomposition/oxidation curve for the Y-based sample indicates such a two-stage, Type II, pyrolysis reaction. X-ray diffraction analysis of the precursor and the product of decomposition immediately after the step in the decomposition curve, however, did not support the idea of the formation of a metastable citrate salt. From the XRD data it was clear that the precursor was not amorphous for an YBCO sample when pH control had not been used. Barium nitrate could always be detected as a crystalline phase, sometimes in association with traces of gerhardtite. Thus it appears that all the higher barium content in the $YBa_2Cu_3O_{7-x}$ superconductor could not be retained in solution during gel production under these conditions. The barium nitrate could be detected up to the end of the step in the decomposition curve and so it is apparent that the form of the pyrolysis curve arises from the presence of barium nitrate and not the production of a metastable mixed citrate salt.

Further evidence of this conclusion is apparent in Fig. 6 where TG data are shown for precursors in which Ho and Eu are substituted for Y in the superconductor and for a precursor in which Eu is substituted for both Y and Ba in the 123 superconductor. In the former cases the step is still present as the result of the presence of Ba whereas in the latter case the step is absent as Eu is substituted for both Y and Ba.

Previous experience using DTA to study the production of oxide powders from citrate precursors has shown that two-stage firing is preferable to single-stage firing as the decomposition of the precursor is a very exothermic process (19). Accordingly, a low temperature firing at 300°C was always carried out in air prior to a higher temperature one at 700°C in oxygen. Figure 7 shows the

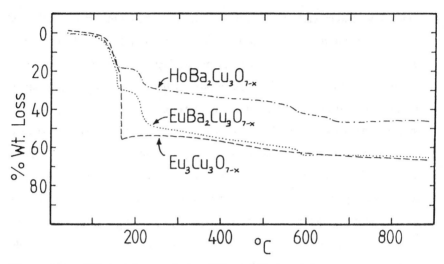

<u>Figure 6</u>:   TG data for pyrolysis of Ho and Eu-based precursors.

dilatometric data for $YBa_2Cu_3O_{7-x}$ produced at 300-700°C and 300-900°C along with data for $HoBa_2Cu_3O_{7-x}$ produced at 300-900°C. These data refer to sintering in air. It is clear that the lower temperature decomposition/oxidation of the precursors has produced powders which start to sinter at lower temperatures. In addition the linear shrinkage percentage is large for the low temperature products of pyrolysis. Both of these effects are explained by the difference in the particle sizes of the products. The finer powders gave a lower pressed density ($\approx 50\%$) than the higher temperature products ($\approx 65\%$). The finer grain size also gives a larger thermodynamic driving force (minimisation of Gibbs surface free energy) for the sintering process so it commences at a lower temperature.

The uptake of oxygen during cooling of sintered materials is of vital importance in order to obtain optimum superconducting properties. Cooling rates of the order of 1.5°C min^{-1} in oxygen have been shown to be highly beneficial in the post-sintering process of the superconductor. Dilatometric data for the sintering and subsequent cooling in oxygen of a sample of $YBa_2Cu_3O_{7-x}$ are shown in Fig. 8. The continual expansion of the specimen on cooling, with the small inflexion in the curve at the temperature of the tetragonal-orthorhombic phase transition, shows the continuous up-take of oxygen during this cooling

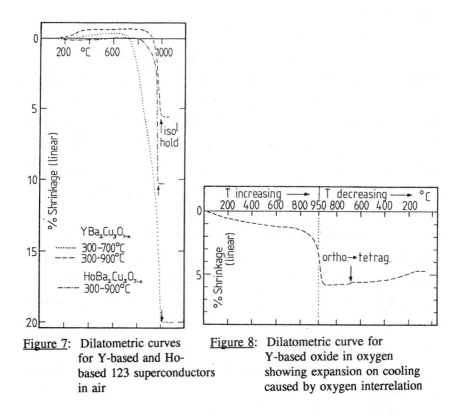

Figure 7: Dilatometric curves for Y-based and Ho-based 123 superconductors in air

Figure 8: Dilatometric curve for Y-based oxide in oxygen showing expansion on cooling caused by oxygen interrelation

cycle. Such dilatometric data agree with oxygen intercalation measurements made using DSC and TG. It is clear from the data presented that TG and dilatometry are vital tools in the study of gel processed electronic ceramics. Today, these techniques are being used with EGA and STA to study the influence of Pb doping on EDTA gel-derived BSCCO superconductors (22).

3. STUDIES OF INTUMESCENT COATINGS

Intumescent coatings are widely used as a form of passive fire protection for structural steel. Intumescence is defined as "to swell up", and this explains the principle action of an intumescent coating, the appearance of which should not differ from any conventional decorative coating (except that it may be applied to a higher film thickness). Active ingredients are incorporated into the coating formulation, such that reaction occurs on exposure to elevated temperature causing the coating to swell to up to 80 times its original volume producing an organic char based layer that insulates the substrate steel.

In Figure 9 below, the solid curve represents the time vs temperature curve for a fast spreading fire as used by the British Standards (23) for the fire testing of commercial intumescent coatings. At this rate of heating, the temperature of an untreated steel column rises according to the dotted line. The critical temperature at which a typical structural steel loses its strength is 550°C. For untreated steel this point is reached in approximately 17 minutes. The coated steel however reaches this point after well over an hour, and with higher film thicknesses fire resistance times of up to 2 and 3 hours are possible.

A typical intumescent formulation contains three active agents, an inorganic pigment such as TiO_2 and a resin binder. The active agents can be characterised as an acid source, a carbonific (or polyhydridic) compound and a spumific agent. These are designed so that the acid source decomposes to yield a mineral acid which dehydrates the carbonific compound to yield a carbon based char. Then at a slightly higher temperature the spumific agent decomposes giving off gases that cause the organic based char to foam and swell out. At higher temperatures the organic layer burns off to yield a finely divided ceramic material based on the TiO_2.

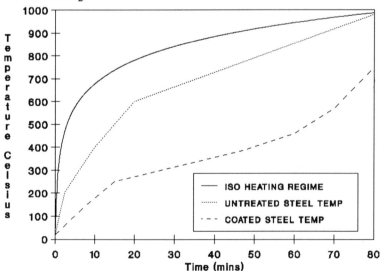

Figure 9:   Time vs temperature curves to illustrate the function of an intumescent coating

The literature published on intumescent coatings lists many possible active agents (24) however the reactions of only one of each class of active agent will be discussed here. The three compounds used are shown below in Figure 10. Knowledge of the thermal stability of each compound and the stabilities of mixtures is vital in understanding the reaction mechanisms involved in intumescent coatings and so thermal analysis may play an important role.

The TG/DTA trace obtained for the thermal analysis of ammonium polyphosphate (APP), which is shown in Figure 11, can be characterised into two stages. The first stage involves a loss of water and ammonia, characterised by EGA as more ammonia than water

AMMONIUM
POLYPHOSPHATE
(ACID SOURCE)

$$-\overset{\overset{\displaystyle O}{\|}}{\underset{\underset{\displaystyle ONH_4}{|}}{P}}-O-\overset{\overset{\displaystyle O}{\|}}{\underset{\underset{\displaystyle ONH_4}{|}}{P}}-O-\overset{\overset{\displaystyle O}{\|}}{\underset{\underset{\displaystyle ONH_4}{|}}{P}}-O-$$

PENTAERYTHRITOL
(CARBONIFIC)

$$C-(CH_2OH)_4$$

MELAMINE
(SPUMIFIC)

Figure 10: The three active agents discussed in this paper.

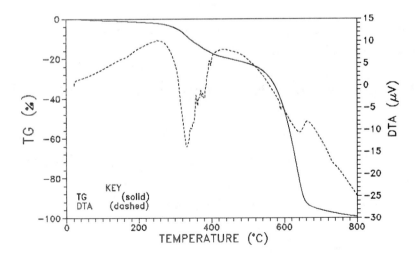

Figure 11: TG/DTA of Ammonium Polyphosphate (heating rate at 10°C min^{-1}, in air at 100 ml min^{-1})

Camino et al (25) have proposed two different cross-linking reaction mechanisms which lead to the formation of phosphate based polymers. These phosphate based polymers break down in the second stage to yield P_2O_5, and cause 100% weight loss from the ammonium polyphosphate.

The carbonific also gives a two stage reaction (Figure 12). The first step is characterised on the DTA by a melting endotherm at 184.7°C. The second stage can be characterised by 100% weight loss from the TG trace, and by either an exotherm if the experiment is carried out in air as shown in Figure 12 or a similar temperature endotherm if the process is carried out in nitrogen. This implies that either a burning or a decomposition process is possible.

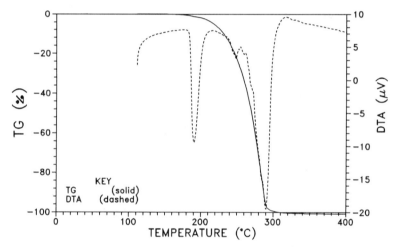

Figure 12:   The TG/DTA trace of pentaerythritol (heating rate 10°C min^{-1}, in nitrogen at 100 ml min^{-1}).

The TG/DTA trace for the spumific, melamine, given below in Figure 13 shows only the sublimation of melamine at 291°C.

Further useful data are obtained from STA studies of the various possible two component systems. When the ammonium polyphosphate and pentaerythritol are mixed together several reactions occur on heating (Figure 14). The first peak can be assigned directly as the PER melting endotherm, however beyond this first stage the trace has little resemblance to either of the two traces corresponding to the pure compounds. The second stage of the reaction occurs from 200°C to approximately 250°C, which is also characterised by an endotherm. Camino et al (25) have suggested that the initial reaction involves either the direct phosphorylation of the PER or alcoholysis of the APP.

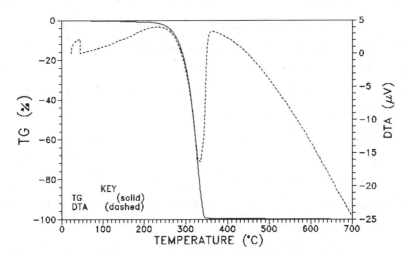

<u>Figure 13</u>:   The TG/DTA trace of melamine (heating rate 10°C min^{-1}, in air 100 ml min^{-1})

A further possibility is the initial loss of ammonia from the APP to yield polyphosphoric acid (PPA). This polyacid then reacts with the pentaerythritol in an esterification reaction prior to a cyclisation reaction.

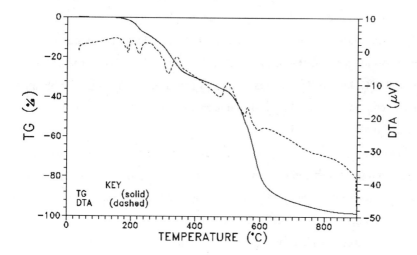

<u>Figure 14</u>:   Ammonium polyphosphate pentaerythritol binary TG/DTA (heating rate 10°C min^{-1}, in air 100 ml min^{-1}).

Because of the low weight loss observed by TG, it can be theorised that the phosphorylation reaction is the most likely. Weight loss in the second step could also be due to the product of the first reaction decomposing to yield ammonia, water and a cyclic intermediate. In the third step, where the resultant DTA trace is very complicated, these APP/PER reactions may occur along with some APP/APP crosslinking reactions give rise to a very complicated random polyphosphate/pentaerythritol crosslinked polymeric structure. Beyond these complicated stages, the TG/DTA traces show a thermal decomposition of the char, as it breaks down to P_2O_5, and CO_2 which are volatilised to give the final weight losses.

The pentaerythritol melamine binary system showed a reaction at a high onset temperature, but then gave no thermally stable products, so has been disregarded with respect to the ternary system.

Data for the ammonium polyphosphate melamine binary system are given in Figure (15) which shows most noticeably 50% weight retained at temperatures as high as 800°C. This thermal stability is much higher than expected, and implies that a reaction has occurred between the APP and the melamine. The complexity of the TG/DTA trace is due to the possibility of APP/melamine reactions similar to those mentioned above as well as APP/APP and also melamine/melamine crosslinking reactions as discovered by May (27). These ultimately give rise to highly complex polyphosphate/melamine polymers.

The TG/DTA trace for the ternary system is shown in Figure 16, where the DTA data resemble those for the APP/pentaerythritol binary at temperatures up to 300°C, but the TG trace is similar to that of the APP/melamine binary at higher temperatures implying that the ternary system involves a mixture of reactions resulting in the formation of random polyphosphate/pentaerythritol/melamine based crosslinked polymeric structures.

In this third example that has been selected it is clear that thermal analysis has enabled an investigation to be made of the chemical mechanisms that occur within commercially available intumescent coating systems. In a system where the temperatures at which chemical events occur is critical, thermal analysis is a very powerful analytical tool.

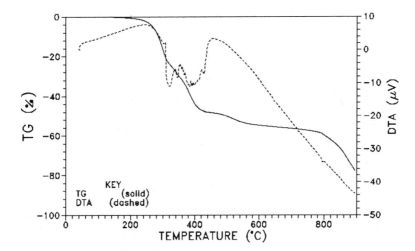

Figure 15: The TG/DTA trace of an ammonium polyphosphate melamine mixture (heating rate 10°C min^{-1}, in air 100 ml min^{-1}).

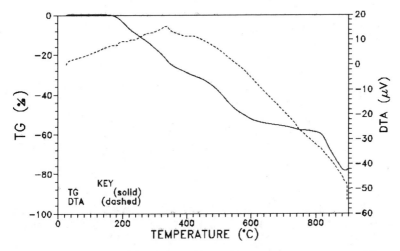

Figure 16: TG/DTA of the ternary mixture (heating rate 10°C min^{-1}, in air 100 ml min^{-1}).

4. CONCLUSIONS

As outlined in the introduction the three areas of work selected for presentation are a minor part of all that may be included in "Materials Science". It is evident that in these, and in many other areas, the importance of thermal analysis cannot be over-emphasised. Surprisingly, relative to studies of polymeric materials, these techniques are not used as much as they could be in the study of metals and ceramics, although their applications are clear. There is no doubt that this is partially because of the difficulties involved in high temperature operations. However, with the increasing reliability and robustness of some modern equipment it is hoped that thermal analysis at elevated temperatures will become an everyday experimental technique of the materials scientist.

REFERENCES

1.   R.H. Beton and E.C. Rollason, J. Inst. Metals, 1958, 86, 77.
2.   A.K. Mukhopadhyay, C.N.J. Tite, H.M. Flower, R.J. Gregson and F.R. Sale, J. de Physique, 1987, 48, C3-439.
3.   H.M. Flower and P.J. Gregson, Mat. Sci. Tech., 1987, 3, 81.
4.   R. Nozato and G. Nakai, Trans. J.I.M., 1977, 18, 679.
5.   J.M. Papazian, C. Sigli and J..M. Sanchez, Scripta Met., 1986 20, 201.
6.   P. Howard, R. Pilkington, G.W. Lorimer and F.R. Sale, High Temp-High Press., 1985, 17, 123.
7.   G. Thomas and J. Nutting, J. Inst. Metals., 1960, 88, 81.
8.   G.W. Lorimer and R.B. Nicholson, Acta Metall., 1966, 14, 1009.
9.   W.H. Hildebrandt, Metall. Trans. Sect. A., 1979, 10, 1045.
10.   F.K. Roehrig and T.R. Wright, J. Vac. Sci. Technol., 1972, 9, 1368.
11.   D.W. Johnson, P.K. Gallagher, F. Shrey and W.W. Rhodes, Ceramic Bull., 1976, 55, 520.
12.   D.J. Anderton and F.R. Sale, Powder Metall., 1979, 1, 8.
13.   F.R. Sale, Metall. and Mat. Technol., Aug., 1977, 439.
14.   J.L. Woodhead and D.L. Segal, Chem. in Britain, April 1984, 310.
15.   J.M. Fletcher and C.J. Hardy, Chem. Ind., 1968, 48.
16.   K.D. Budd, S.K. Dey and D. Payne, British Ceramic Proceedings, 1985, 36, p.107.
17.   D.J. Anderton and F.R. Sale, Powder Metall., 1979, 1, 14.
18.   M.S. Baythoun and F.R. Sale, J. Mater. Sci., 1982, 17, 2757.
19.   F. Mahloojchi, F.R. Sale, High Tech Ceramics, 1986, Elsevier.
20.   F.R. Sale and F. Mahloojchi, Ceramics Int., 1988, 14, 229.
21.   P.H. Courty, H. Ajot and C.H. Marcilly, Powder Technol., 1973, 7, 21.

22.   M. Rajabi and F.R. Sale, in Euro-Ceramics Vol. 2, Properties of Ceramics, Ed. G. de With, R.A. Terpstra and R. Metselaar, 1989, Elsevier Applied Science, London and New York, p. 2426.
23.   British Standards Institute BS476, part 20.
24.   H.L. Vandersall, J. Fire and Flamm., 1971, 2, 97.
25.   G. Camino, L. Costa and L. Trossarelli, Polym. Deg. Stab., 1985, 12, 203.
26.   G. Camino, L. Costa and L. Trossarelli, Polym. Deg. Stab., 1985, 12, 213.
27.   H. May, J. Appl. Chem., 1959, 9, 340.

Minerals to Fossil Fuels

S. St. J. Warne

DEPARTMENT OF GEOLOGY, THE UNIVERSITY OF NEWCASTLE, SHORTLAND
NSW 2308, AUSTRALIA

1   INTRODUCTION

The rapid expansion in the scope, increasing number of
thermal analysis (TA) methods available, and the wide
range of applications which in recent years have been
documented has led to the recognition of TA as a branch
of science in its own right as instanced by its vital and
expanding role in the polymer industries of today.[1]

Mirroring this is a lesser known but important
resurgence in utilization in the earth sciences, where
applications to a wide range of topics are continuing to
rapidly expand.[2] Examples of areas of application
range from minerals, mineral mixtures, members of isomor-
phous substitution series, soils, ceramics, cements, con-
crete aggregates, building stones, natural materials
characterization e.g. industrial raw to end product materials assessment, together with their thermal stability,
performance purity and quality control.[2]

Contrasting sets of applications are in the areas of
fossil fuels and associated environmental aspects.[3]
Here, as coal and oil shale uses invariably involve
heating, thermal analysis methods are particularly
suitable for many aspects of fundamental and applied
research i.e. for their assessment, utilization,
extraction and end product re-use and or characteristics
prior to disposal.

In more generally applicable ways TA may be applied
to the earth sciences to investigate thermal stability,

physical changes, reactivity, decomposition and recombination reactions. Also, to characterize and identify
reactions, for identification, evaluation, industrial
performance and quality control purposes, the mineral
contents of mixtures and their detection limits.[3]

2   AIMS AND APPROACH

As the above indication of earth science topics is so
large it would require a multi-volume series to cover it
adequately, an up to date overview approach has been adopted as detailed below. In this way through a considerable number of suitable references the title subject,
"Minerals to Fossil Fuels" has been introduced and some
highlights reviewed.

This has been achieved by indicating the range of
areas of application supported by comprehensive book
coverages together with indications of the types of chapters from books not restricted to the mineralogical or
fuel topics together with examples of reviews. To complement this, a selection of recent applications using, in
particular, modern methods and techniques has been added.
In an integrated way it has therefore been the overall
aim:-

(1)   To provide a suitable background entree into the
       subject for scientists newly exposed to TA in the
       earth sciences.

(2)   Further, to indicate something of the wide scope
       of work involved in the current resurgence of
       applications to natural (geological) materials.

(3)   To illustrate a sample of industrial and environ-
       mental topics which today are of ever increasing
       importance.

(4)   Finally, to draw attention to a newly developed
       earth science method which also shows very consid-
       erable potential promise in the general field of
       TA viz. Proton Magnetic Resonance Thermal Analysis
       (see below).

3 MINERALOGY

Background

Comprehensive accounts of the early applications of TA have recently appeared in the fascinating work "A History of Thermal Analysis".[4] From this there seems little doubt that the work of Henry Le Chatelier which culminated in 1887 in the development of his ingenious automated TA unit and its applications to minerals[5,6] paved the way for the applications of TA in the earth sciences.

There followed a rapid and somewhat undisciplined expansion of applications to mineralogy which were systematized and consolidated in several publications by Robert Mackenzie, which were initiated by the "Scifax Differential Thermal Analysis Data Index"[7] and led, in 1970 and 1972, to the twin ageless companion volumes titled "Differential Thermal Analysis",[8] which continue as invaluable references. Subsequently a number of complementary books,[9-17] specific chapters in reference books[18-20] and reviews,[2,4,21-25] have led collectively and individually to the current resurgence in applications in the earth sciences.

4 SOLID FOSSIL FUELS

Background

Solid hydrocarbon rich fossil fuels (coal and oil shale) are of sedimentary origin and as such are composed of two fundamentally different groups of constituents[3] viz.

(1) The organic (hydrocarbon bearing) macerals, which are derived from the plant materials which were present in the original peat swamps.

(2) The inorganics (minerals) which were blown, washed or chemically precipitated into the peat.

Thermal analysis methods have been widely used in the general areas of coal seam evaluation, washability

(beneficiation), utilization, waste and final product
assessment, petrochemical aspects such as pyrolysis and
kinetic studies and environmental considerations such as
spontaneous combustion and acid rain.

For these reasons fossil fuel research provides an
excellent example of the modern applications of TA in a
specialized branch of the earth sciences today.

5   GENERAL APPLICATIONS

Thermal analysis applications in the earth sciences have
involved the identification of specific reactions or sets
of reactions, which are endothermic/exothermic or measure
other properties which are:-

   (1)   Characterized by weight variations which in turn
         are associated with the evolution of, and or
         recombination with, one or more gases (calcite
         [8,18,30], shutterudite[26] and caledonite[27]).

   (2)   The result of crystallographic inversions/revers-
         ions (quartz[8,11]), melting/recrystallization[11]
         and decomposition, dissociation or solid
         state reactions not involving gases, which are
         characterized by no weight changes at all.[28]

   (3)   The result of the variation in other properties
         which can be measured by different TA methods e.g.
         thermomagnetometry (TM),[23] thermomechanical
         analysis (TMA),[29] thermoluminescence (TL),[17]
         emanation thermal analysis (ETA)[30] or thermo-
         microscopy.

Simultaneous Thermal Analysis

   Simultaneous DTA-TG not only further quantifies the
endothermic/exothermic DTA reactions by measuring the
weight variations involved,[31] but also indicates those
reactions which take place at constant weight.[27]

   Similarly simultaneous DTA-EGA[32] (Figure 1), TG-
EGA or DTA-TG-EGA[27] (with the EGA being determined by
IR, GC or MS), aids in the quantification of the DTA
reaction peaks in a different way by identifying the gas

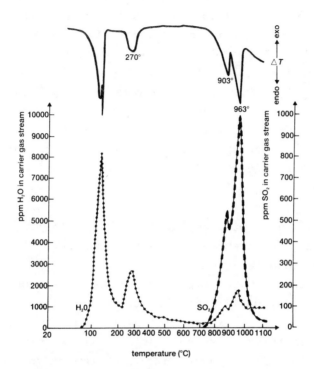

<u>Figure 1</u> Simultaneous DTA/EGA curves of $CuSO_4.5H_2O$ using non-dispersive infra-red analysers for determining H_2O and SO_2 evolution during heating. (After Morgan 1977, reprinted with permission of the Journal of Thermal Analysis).

given off and its amount.[33]

A fundamental advantage of simultaneous application of TA methods is that they are applied at the same time to a single sample under exactly identical conditions of analysis with either one TA unit or closely linked units.

<u>Diagnostic Mineralogy</u>

By whatever method or technique used, the thermal analysis curves obtained from, in particular, DTA and DSC and to a somewhat lesser degree TG and EGA, are very often of diagnostic mineralogical value. This applies to single minerals[7,11] and minerals in mixtures.[34]

On the other hand the identity of the mineral, rock

or their directly derived products, as used in industry,
may be known and it is the types of reaction these undergo
on heating[35] their degree of reactivity,[36] plasticity,[37] or
thermal stability[38] which is of importance.

Further, the effects on these reactions of impurities[39] or additives,[40] catalysts or simply the presence
of other minerals in natural or artificially beneficiated
or prepared mixtures can cause additional thermal effects,
which if carefully investigated may be of further
diagnostic, industrial or research value.[2]

Reaction Types

The types of reactions applicable in mineralogy,
which may be detected, using for example DTA, may be
divided into two groups.

(1)   Endothermic; typified by, decomposition, water
      loss, solid state reactions, reduction, melting
      and crystallographic inversions.

(2)   Exothermic; recombination, mineral/mineral react-
      ions to form new "minerals", oxidation, solidific-
      ation (recrystallization) and crystallographic
      reversions.

The specific identification of such reaction types,
is particularly facilitated by the application of simultaneous and or coupled TA methods, which for example
indicate associated weight losses (TG), resolve superimposed features (TG/DTG)[13] (Figure 2) and or identify
gases which may be evolved (EGA) from the individual
reactions.

In contrast are the methods, also used in simultaneous determinations such as TM[23] or TMA,[41] or even
simultaneous TM/Dilatometry.[42] These measure changes in
the material under test which do not involve the loss of
material but merely changes taking place in it.

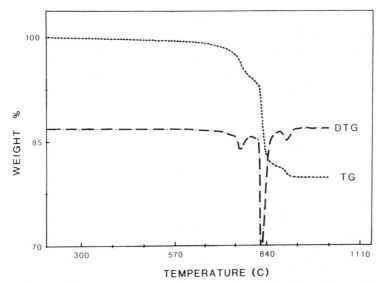

Figure 2 Simultaneous TG/DTG curves of the clay mineral hectorite in a flowing atmosphere of CO_2. (After Earnest 1981, reprinted with permission of Thermochimica Acta).

6 MINERALS IN MIXTURES

Derivative Thermogravimetry

Always, when dealing with mixtures the partial superposition of the TA effects of reactions which take place at similar temperatures is a problem. By the use of TG/DTG separation of these features is facilitated (Figure 2).

Variable Atmosphere Thermal Analysis

The rapidly expanding use of the technique of "variable atmosphere thermal analysis", has recently been reviewed[45] and shown to have a major role to play in applied mineralogy. It involves the preselection, control, maintenance and or change of the furnace atmosphere composition surrounding the sample under test, between or during individual TA runs.

This is achieved under flowing gas (purge) conditions through or over the "unknown". However, although the former theoretically should give the better control, the tendency for the powdered sample material to be disturbed and blown out of the sample holder means that the gas is almost always passed over the sample.[45]

The required gas atmosphere is maintained with the use of individual, mixtures or combinations of different purge gases, singly or in sequence. These may be maintained at ambient, above ambient or vacuum pressure conditions and are equally applicable under dynamic, isothermal or quasi-isothermal heating conditions.

Considerable care must be exercised to ensure the purity of such gases. In particular, commercial "high quality oxyfree nitrogen" far too often contains suffic- ient oxygen to prohibit it maintaining inert conditions. This is particularly so as it is constantly used under purging conditions. It should therefore always be pretested or passed through an oxygen trap extraction system.

It is vital that the TA system is cleared of one gas before the next is used. This may be achieved by main- taining a period of isothermal conditions during which purging with an inert gas is completed before continuing with heating in the next gas atmosphere to be used.[37]

This aspect may be greatly assisted by restricting the space around the sample as much as possible, so that the previous gas may be removed quickly and effectively e.g. the enclosing small micro-environment cup which is an integral, part of the Stanton Redcroft STA thermal analysis units.[46]

It is also important to note that DTA or DSC peak areas and temperatures obtained from identical samples, will show variations when determined in inert gases of different thermal conductivity. See details in chapter "Introduction to Thermal Analysis Methods" herein.

The preferential movement of specific reactions, up or down scale, due to increased or decreased purge gas partial pressures, which leaves other types of reactions unaffected, provides another way of resolving superim- posed decomposition reactions.[34] This technique is particularly applicable to DTA, DSC, TG and to a lesser degree TM.

Double Differential Thermal Analysis

A useful technique applicable to the DTA of mineral mixtures and the more accurate determination of their specific mineral contents, is double differential thermal analysis.[43] This involves adding to the inert reference, what is considered to be an equal amount of one of the suspected minerals present in the unknown sample.

On DTA the peak or peaks of this mineral (if it is the correct one and in about the same amount) will be nullified and be greatly reduced or caused to "disappear" from the resultant DTA curve. The presence of this mineral is thus confirmed and the remaining peaks which represent the other mineral components in the mixture are left free from interference.

Conversely if there is no diminution of the original peaks and new peaks are added, then the mineral added to the inert reference is clearly not one of those present in the original mixture under test.

In addition this technique has useful applications for individual mineral content evaluations when present in mixtures. Thus the amount added to the reference to cause complete nullification is equal to the amount of this mineral in the unknown sample. Complementary to this is the contribution it can make to the resolution and identification of the components of superimposed peaks. This is achieved by the removal of the peak of one component leaving the other clearly defined without interference from a simultaneously occurring reaction.

Differential Reaction Analysis

Another valuable, but surprisingly little known technique is termed "differential reaction analysis"[44] and has been developed to obtain DTA from a number of minerals (some common) which are thermally inert in the traditional range of TA investigation, i.e. ambient to about 1000°C.

The procedure is to react two substances i.e. the nonreactive material with a suitable chemical reagent. The resultant material to be applicable will decompose on DTA to give a curve of diagnostic value. The published account[44] gives details of the minerals found suitable for recognition by this method together with the reactants used.

Simultaneous Thermal Analysis

The simultaneous application of at least two TA methods to the same sample under identical conditions of analysis provides for the detection of parameters which may be produced from one mineral in a mixture and not another.

In this way specific reaction types characteristic of individual minerals, or mineral groups, present may be recognised, identified and measured against simultaneously determined reference curves produced for example by DTA and or TG.

For example, if the simultaneously determined DTA/EGA of a sample exhibits on its DTA curve a large endothermic peak, but the EGA curve shows only 50% of the evolved carbon dioxide gas this should represent, then an anomaly has been detected.

The explanation could be that a second reaction is taking place at the same time, which for example, liberates a different gas, perhaps water. Clearly, if the required composite weight loss could be confirmed from a simultaneously determined TG curve i.e. by DTA/TG/EGA, while specific gas detectors for carbon dioxide and water confirmed the production of these gases, then the presence of more than one mineral would be indicated.

In this way carbonate, sulphide and clay minerals typically give off only one gas viz. carbon dioxide, sulphur dioxide or water vapour respectively.

Conversely other minerals may exhibit different properties such as magnetism (TM), calorific values (DSC) or thermoluminesence (TL) at lower than expected intensities, which are indicative of components in mixtures and not single minerals.

7 CONTRASTING EXAMPLES OF APPLICATIONS OF VARIABLE ATMOSPHERE THERMAL ANALYSIS

A wide range of recent and current applications of TA in the earth sciences involves the use of pre-selected controlled furnace atmosphere conditions. For this reason this technique has been concentrated on herein because it presents a good indication of the role of TA in the earth sciences today. Further, the renewed interest in this area is such that some applications have lately become routine practice.[15]

General Applications

The applications of this technique fall into several groups. These basically depend on the presence, absence or partial pressure of a particular gas in the furnace atmosphere affecting chemical reactions which show as modifications on the resultant TA curve[45] and may well result in the following which are best illustrated using DTA as a model (Figure 3).

(1) The suppression or enhancement of individual reactions.

(2) The identification of specific reaction types by their preferential upscale movement under increased partial pressure conditions of the right gas e.g. carbonate peaks, in flowing carbon dioxide, but not sulphide peaks liberating sulphur dioxide, when determined in the same conditions of flowing carbon dioxide.[37]

(3) The improved detection limits of suitable minerals due to increases in peak heights when determinations are made in higher partial pressures of the reaction gas being evolved. In such cases the delayed decomposition reactions take place more vigorously at a higher temperature to give

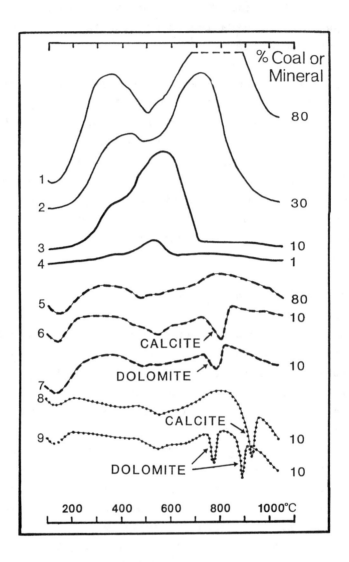

Figure 3 DTA curves of coal diluted with Al_2O_3 (curves 1–5) or 10% calcite or dolomite (curves 6–9), determined in static air (_____), flowing N_2 (————) or flowing CO_2 (......), to show coal combustion or detection limits (curves 1–4), suppressed burning effects in N_2 (curve 5) and clearly defined modifications, to the coal curve, due to the presence of calcite or dolomite (curves 6–7) and the improved characterization of these two minerals in CO_2 (curves 8–9). After Warne 1991, reprinted with permission of Springer Verlag).

increased peak heights.

(4) The resolution of superimposed features by preferential movement of one peak type, but not the others, as per point 2 immediately above.

(5) The identification of specific minerals or mineral groups characterized by such gas atmosphere controlled reactions.

(6) The comparable application of points 1-5 above to additional minerals which give off any other gas if determined in increased partial pressure conditions of this same gas, i.e. water vapour or sulphur dioxide.

(7) The assessment of organic matter contents when in mixtures with minerals.

Specific Application Examples

The use of flowing air or oxygen to promote oxidation has proved valuable for studies of combustion,[48] the detection of small amounts of organic material with minerals,[49] mineral identification,[50] magnetic component detection and coal mineral/ash contents.[51]

Conversely inert conditions using nitrogen, argon, helium and sometimes carbon dioxide, have been applied to prohibit oxidation and burning and remove the effects of the partial pressures of other gases (e.g. carbon dioxide in air).

With these interfering or superimposed effects removed much improved specific reaction definition, mineral identification, detection limits, and for coal its mineral contents, volatile yields, ash constituents and liquefaction residue determinations have resulted.

Peak suppression, using inert gases has been applied in DTA to remove the vigorous, broad exothermic oxidation reactions which often completely or partially overprint smaller endothermic reactions taking place within the same temperature range.

For example the suppression of coal combustion (strongly exothermic), allows the much smaller endothermic peaks of the minerals present in coal samples to be recorded[3] or the endothermic decomposition of siderite (iron carbonate) to form FeO. This in air, due to its very high oxidation potential, oxidizes exothermically to hematite so rapidly that it takes place before all the siderite has decomposed.

Thus, these two reactions are virtually superimposed on the resultant DTA curve where the size of the endothermic decomposition peak is partly or completely negated and therefore does not correctly represent the original siderite content. However, in an inert atmosphere no oxidation takes place leaving the complete endothermic decomposition peak to accurately represent the amount of siderite present.[47]

For coal and oil shales a detailed list has been presented specifically for DTA and DSC in the comprehensive review by Warne and Dubrawski.[52] Examples of wider applications,[3] range from organic contents,[49] coal rank,[53] mineral constituents and contents,[18] combustion characteristics[48,54] and calorific values [55,56], proximate analysis,[57] sulphur contents,[58,59] coal quality and ash assessment,[60] quality control[49-60] and coking potential.[35]

Environmental aspects relate to unwanted products such as bottom and fly ash, slag, coal washery wastes and the production of the gaseous oxides of sulphur, nitrogen and carbon and their links to acid rain and the greenhouse effect.[3]

The physical and chemical stability of these solid products, together with their reactivity, and the huge amounts produced, presents major negative economic costs for physical transport and disposal. Further, stockpile spontaneous combustion and dump leachate waters may have toxic considerations involving heavy metal enrichment or have unacceptable pH which may sometimes be related to the original minerals present.[3]

8 PROGRAMMED TA USING DIFFERENT GAS SEQUENCES

In the environmentally related area several recent rather complex TG technique manipulations have provided valuable results.[15,60]

As these not only break new ground, but clearly indicate considerable potential for future applications in other areas of TA, they have been included here as a "state of the art" conclusion to this section concerning aspects of applications of TA to earth science materials.

This group of related and progressively more complex applications was initiated by the development of a most successful TG method for the proximate analysis of coal.[57] It involved a multi-stage heating programme accompanied by furnace gas atmosphere changes during the course of individual TG runs of coal to determine the four required proximate analysis parameters, i.e. moisture, volatile, fixed carbon and ash contents.

Nitrogen-Oxygen-Nitrogen Regime

From Figure 4, it is clear, that in inert conditions

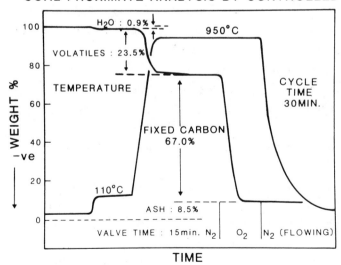

COAL PROXIMATE ANALYSIS BY CONTROLLED TG

Figure 4   TG applied to the proximate analysis of coal. (After Earnest and Fyans 1981, reprinted with permission of Perkin Elmer).

of flowing nitrogen the coal sample is rapidly heated to
110°C where it is held to constant weight to give the
first weight loss due to the moisture content. Still in
flowing nitrogen the same sample is rapidly heated to
950°C and again held to constant weight to produce the
second weight loss due to the evolution of volatiles. At
this point the purge gas is changed from nitrogen to
oxygen and burning takes place to provide the third weight
loss due to the combustion of the residual (fixed) carbon.
The remaining residual mass finally determined with a
change back to flowing nitrogen, for accurate comparison
with the original coal mass, is the ash content which
represents the original mineral matter in the coal.[57]

Nitrogen-Oxygen-Nitrogen-Hydrogen Regime

In this case the cold coal ash from a proximate
analysis determination described above using the first
three gases is now subject to TM by placing a fixed
magnet above the sample which is reheated in flowing
hydrogen. Under these conditions the amount of iron
oxide in the ash which came from the decomposition of any
pyrite (iron sulphide) in the original coal sample is
reduced to metallic iron. This iron content, due to its
attraction by the magnet above, shows as an apparent
weight gain on the resultant TM curve. The magnitude of
this "weight gain" is directly related to the original
pyrite content and provides a measure of the pyritic
sulphur content of the coal.[51]

Air-Air+Sulphur Dioxide-Nitrogen-Nitrogen+Hydrogen-Nitrogen-Air Regime

The ultimate in furnace atmosphere control
complexity may have been reached with the development of
the "sulphation, regeneration and oxidation", (SOR) test.
This has been developed for the testing, as sorbents, of
calcium aluminates and titanates prepared by different
methods.[37]

With reference to Figure 5, it can be seen that the

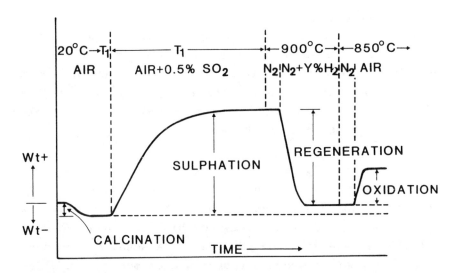

Figure 5 Multi-gas TG for the sulphation, regeneration and oxidation (SRO) test for flue gas sorbents. The weight loss in air is due to calcination, followed in air+SO$_2$ by the large weight gain of sulphation, followed in N$_2$+H$_2$ by the weight loss caused by sorbent regeneration. After N$_2$ purging the final weight gain in air is due to oxidation to form CaSO$_4$. (After Duisterwinkel et al 1989, reprinted with permission of Thermochimica Acta).

initial heating in air gives a weight loss due to calcination. This is followed by heating to a suitable reaction temperature (T1), which often varies for different sorbents, in flowing air plus sulphur dioxide. Here the sample is held isothermally to produce a weight gain due to sulphation of the sorbent. At constant weight the temperature is raised to 900°C where a sequence of different flowing gases is used i.e. first nitrogen, to purge the system, followed by flowing nitrogen + hydrogen in which the weight loss indicates the degree of regeneration exhibited by each different sorbent. Finally to complete this test the sample is cooled to 850°C and held there isothermally, where a second purge of nitrogen is followed by a further change to flowing air. This results in a rapid weight gain due to the oxidation of the CaS formed during regeneration which accounts for the incomplete regeneration of this range of sorbents.[37]

Further examples

The application of other gas sequences to TG runs such as nitrogen-oxygen-carbon dioxide to pre-heated coal, bed ash, fly ash and fluidized bed ash[61] and nitrogen-oxygen to municipal solid refuse derived fuels[62] and the efficiency of fluidized-bed coal combustion[63] have also made important contributions.

9   LITERATURE SOURCES

It is also noteworthy that publication of geoscience TA papers is in no way restricted to the two international journals devoted exclusively to this topic e.g. Journal of Thermal Analysis and Thermochimica Acta. A considerable number of excellent papers also occur in a wide range of other, often specialized journals e.g. the Mineralogical Magazine, Neus Jahrbuch, Clay Minerals, the American Ceramic Society, Fuel, and individual Special Publications such as those by ASTM,[15] the Clay Minerals Society[16] the Mineralogical Society[64] and sometimes individual Companies.[65,66] In respect to this the TA reviews in general and particularly the comprehensive ones which appear regularly with the suffix R in the journal Analytical Chemistry[21] also provide an invaluable source of less well known to virtually unknown publications.

10   REFERENCES

1.   E.A.Turi (Ed.), 'Thermal Characterization of Polymeric Materials', Acad. Press, New York, 1981.
2.   S.St.J. Warne, Thermochim. Acta, 1991, (in press).
3.   S.St.J. Warne, Thermochim. Acta, 1990, 166, 343.
4.   R.C. Mackenzie, Thermochim. Acta, 1984, 73, 251.
5.   H.Le Chatelier, C.R. Acad. Sci., Paris, 1887, 104, 1443.
6.   H.Le Chatelier, Bull. Soc. Miner. Crist. 1887, 10, 204.
7.   R.C. Mackenzie, 'Scifax Differential Thermal Analysis Data Index', Cleaver-Hume, London, 1962 and Supplement 1964.
8.   R.C. Mackenzie, (Ed.), 'Differential Thermal Analysis', Vol.1, 1970 & Vol.2, 1972, Acad. Press, New York.
9.   P.D. Garn, 'Thermoanalytical Methods of Investigation', Acad. Press, London, 1965.

10. D. Schultz, 'Differentialthermoanalyse', VEB, Deutscher Verlag Der Wiessenschaften, Berlin, 1971.
11. W. Smykatz-Kloss, 'Differential thermal Analysis - Application and Results in Mineralogy', Springer-Verlag, Berlin, 1974.
12. D.N. Todor, 'Thermal Analysis of Minerals', Abacus Press, Tunbridge Wells, 1976.
13. C.M. Earnest. 'Thermal Analysis of Clays, Minerals and Coals', Perkin-Elmer Co., Norwalk, 1984.
14. W.W. Wendlandt. 'Thermal Analysis', (3rd. Edit.), John Wiley and Son, New York, 1985.
15. C.M. Earnest (Ed.), 'Compositional Analysis by Thermogravimetry', ASTM Spec. Publ. 997, Philadelphia, 1988.
16. J.W. Stucki, D.L. Bish and F.A. Mumpton (Eds.), 'Thermal Analysis in Clay Science', CMS Workshop Ser. 3, The Clay Minerals Society, Boulder, Colorado, 1990
17. S.W.S. McKeever, 'Thermoluminescence of Solids', Cambridge University Press, Cambridge, UK, 1985.
18. S.St.J. Warne, 'Analytical Methods for Coal and Coal Products', Clarence Karr Jr. (Ed.), Acad. Press, London, 1979, Vol.3, Chapter 52, p.447.
19. N.L. Viona and D.N. Todor, 'Analytical Methods for Coal and Coal Products', Clarence Karr Jr. (Ed.), Acad Press, London, 1978, Vol.2, Chapter 37, p.619.
20. C.M. Earnest, in 'Compositional Analysis by Thermogravimetry', C.M.Earnest (Ed.), ASTM Spec. Publ. 997, Philadelphia, 1988, p.272.
21. W.W. Wendlandt, Anal. Chem., 1984, 56, 250R.
22. D. Dollimore, Anal. Chem., 1990, 62, 44R.
23. S.St.J. Warne, H.J. Hurst and W.I. Stuart, 'Applications of Thermomagnetometry in Mineralogy, Metallurgy and Geology', (invited review), Therm. Anal. Abs., 1988, 17, 1.
24. D.J. Morgan, S.B. Warrington, and S.St.J. Warne, Thermochim. Acta, 1988, 135, 207.
25. J.V. Dubrawski, in, 'Thermal Analysis in the Geosciences', W. Smykatz-Kloss and S.St.J. Warne (Eds.), Springer Verlag, Berlin, 1991, 16.
26. L.J. Wilson and S.A. Mikhail, Thermochim. Acta, 1989, 156, 107.
27. D.J. Morgan, S.St.J. Warne, S.B. Warrington and P.H.A. Nancarrow, Min. Mag., 1986, 50, 521.
28. M.E. Brown, 'Introduction to Thermal Analysis', Chapman and Hall, London, 1988.
29. R.E. Wetton, R.D.L. Marsh and J.G. Van de Velde, Thermochim. Acta, 1991, 175, 1.
30. V. Balek and J. Tolgyessy, 'Emanation Thermal Analysis', Elsevier, Amsterdam, 1984.
31. Y. Deutsch and L. Heller-Kallai, Thermochim. Acta, 1991, 182, 77.
32. D.J. Morgan, J. Thermal. Anal., 1977, 12, 245.
33. A.E. Milodowski and D.J. Morgan, Proc. 6th. ICTA, W.Hemminger (Ed.), Bayreuth, Birkhauser Verlag, Stuttgart, 1980, Vol. 2, p.289.

34. S.St.J. Warne and D.H. French, Thermochim. Acta, 1984, 79, 131.
35. S.St.J. Warne and J.V. Dubrawski, J. Thermal Anal., 1988, 33, 435.
36. R. Sakurovs, L.J. Lynch, D.S. Webster and T.P. Maher, Jour. Coal Qual., 1991, 10, 37.
37. A.E. Duisterwinkel, E.B.M. Doesburg and G. Hakvoort, Thermochim. Acta, 1989, 141, 51.
38. S.St.J. Warne, Thermochim. Acta, 1984, 75, 139.
39. W.R. Bandi and G. Krapf, Thermochim. Acta, 1976, 14, 221.
40. V.R. Choudhary, S.G. Pataskar, M.Y. Pandit and V.G. Gunjikar, Thermochim. Acta, 1991, 180, 69.
41. R.E. Wetton, R.D.L. Marsh and J.G. Van de Velde, Thermochim. Acta, 1991, 175, 1.
42. E. Karmazsin, P. Satre and P. Vergnon, J. Thermal. Anal., 1983, 28, 279.
43. P.D. Garn, 'Thermoanalytical Methods of Investigation', Acad. Press, New York, 1965, p.606.
44. Z.H. Zuberi and O.C. Kopp, Amer. Mineral., 1976, 61, 281.
45. S.St.J.Warne, in, 'Thermal Analysis in the Geosciences', W. Smykatz-Kloss and S.St.J. Warne (Eds.), Springer Verlag, Berlin, 1991, 62.
46. E.L. Charsley, J. Joannou, A.C.F. Kamp, M.R. Ottaway and J.P. Redfern, Proc. 6th. ICTA Conf., Bayreuth, (W.Hemminger Ed.), Birkhauser Verlag, Stuttgart, Vol.2, 1980, p.237.
47. S.St.J. Warne, Thermochim. Acta, 1987, 110, 501.
48. J.W. Cumming and J. McLaughlin, Thermochim. Acta, 1982, 57, 253.
49. S.St.J. Warne, Thermochim. Acta, 1985, 86, 337.
50. S.St.J. Warne, Chemie der Erde, 1976, 35, 251.
51. D.M. Aylmer and M.W. Rowe, Thermochim. Acta, 1984, 78, 81.
52. S.St.J. Warne and J.V.Dubrawski, J. Therm. Anal., 1989, 63, 219.
53. Setaram, 'Fossil Fuels Application Sheets 1-4', Lyon, undated.
54. D.E. Rogers and M.D. Biddy, Thermochim. Acta, 1979, 30, 303.
55. C.M. Earnest, Instrum. Res., 1985, 1, 73.
56. S.St.J. Warne and J.V. Dubrawski, J. Therm. Anal., 1988, 33, 435.
57. C.M. Earnest and R.L. Fyans, 'Recent Advances in Microcomputer Controlled Thermogravimetry of Coal and Coal Products', Perkin-Elmer Thermal Analysis Application Study No.32, 1981.
58. D.M. Aylmer and M.W. Rowe, Thermochim. Acta, 1984, 78, 81.
59. S.St.J. Warne, A.J. Bloodworth and D.J. Morgan, Thermochim. Acta, 1985, 93, 745.
60. S.St.J. Warne, Trends, Anal. Chem., 1991, 10, 195.
61. C.F. Culmo and R.L. Fyans, in 'Compositional Analysis by Thermogravimetry', C.M.Earnest (Ed.), ASTM Spec. Publ. 997, Philadelphia, 1988, p.245.

62.   R.K. Agrawal, in 'Compositional Analysis by
      Thermogravimetry', C.M.Earnest (Ed.), ASTM Spec.
      Publ., 997, Philadelphia, 1988, p.259.
63.   S.A. Mikhail and A.M. Turcotte, <u>Thermochim. Acta</u>,
      1990, <u>166</u>, 357.
64.   R.C. Mackenzie, 'The Differential Thermal Anlaysis
      of Clays', Mineralogical Society, London, UK, 1957.
65.   C.M. Earnest, 'Thermal Analysis of Clays, Minerals
      and Coal', Perkin Elmer Corp., Norwalk, USA, 1984.
66.   G. Widmann and R. Riessen, 'Thermal Analysis:
      Terms, Methods, Applications', Huthig, Heidelberg,
      1987.

Catalysts—Thermal Analysis Applications

D. Dollimore

DEPARTMENT OF CHEMISTRY AND COLLEGE OF PHARMACY, UNIVERSITY OF TOLEDO, TOLEDO, OH USA

1 DEFINITION AND CHARACTERISTIC FEATURES OF A CATALYST

In text books *catalysts* are defined as materials which accelerate a chemical reaction by their presence but which are unchanged chemically in the process.[1] The process is called *catalysis*. In some older textbooks a catalyst is defined as a material which would alter the rate of a chemical reaction without itself undergoing a chemical change. The term *negative catalyst* was used to indicate a catalyst which slowed down a reaction whilst a catalyst which accelerated a reaction would be termed positive. The term *inhibitor* has come to be used as a replacement for "negative catalyst" but the manner of its use would cover materials which may be consumed during the reaction. Catalysis can be divided up into two broad classes, homogeneous catalysis and heterogeneous catalysis. In homogeneous catalysis the catalyst is part of the same phase as the reactants. In heterogeneous catalysis the catalyst forms a separate phase. Now it can be demonstrated that in heterogeneous catalysis there may be small changes in the stoichiometry of the catalyst which renders the text book definition inadequate. The same may be true of homogeneous catalysts but the problem in the latter class is generally the recovery of the catalyst so that the product may be described as pure. In modern thermogravimetric instrumentation the equipment is generally sensitive enough to follow changes in the stoichiometry of heterogeneous catalysts. It may be noted that the failure of catalysts like manganese dioxide to conform to the exact stoichiometry MnO_2 and its change during the catalyst process would have made Berthelot a happy man, and if his theories about non-stoichiometry had been accepted by Dalton the development of the Atomic Theory might have progressed along somewhat different lines.

The application of Thermal Analysis (TA) to catalysts, catalysis, and chemisorption have been mainly in the field of heterogeneous catalysis. The two main thermal analysis techniques employed to study catalysts are thermogravimetry (TG) and differential thermal analysis (DTA). However as much of the thermal analysis work is now quantitative following the more recent conventional use, the term differential scanning calorimetry (DSC) should be used in such circumstances.

It will be noted that a third term has been introduced namely chemisorption and it is quite impossible to continue any discussion on catalysis without using this term. Chemisorption is the process of adsorption but is distinct from van der Waals or physical adsorption in that a chemical bond is formed between the adsorbate and the adsorbent surface. This process of chemisorption and the description of the chemisorbed species is at the heart of catalysis process. Most often the chemisorbed species is only lost from the surface by raising the temperature when material is degassed as a different kind of species. Thus oxygen chemisorbed on carbon is desorbed as carbon dioxide and carbon monoxide mixtures. This forms the basis for the technique of thermal desorption which may be regarded as a special technique related to Thermal Analysis.[2] In this technique the desorption of various chemisorbed species may be studied either at constant temperature against time or more usually against temperature as the system is subjected to a rising temperature regime.

The general characteristics of catalysis are often quoted as:

1.   "The catalyst is unchanged chemically at the end of the reaction."
2.   "A small quantity of catalyst is enough to affect the rate of reaction."
3.   "The catalyst does not affect the equilibrium in a reversible reaction."
4.   "The catalyst does not initiate the reaction – it can only alter the rate of a reaction already possible."

The use of catalysts is important industrially, e.g. the combination of nitrogen and hydrogen to form ammonia, the oxidation of ammonia, the oxidation of carbon monoxide in the presence of air, the Fischer-Tropsch reaction, the shift reaction, and the oxidation of sulfur dioxide to sulfur trioxide.

In heterogeneous catalysis utilizing a solid phase catalyst the envisaged process is that at the surface of the catalyst the reactant phases are adsorbed to form a new species which is subsequently desorbed as a gaseous product. In such a process, the surface atoms or ions are often involved. The catalytic solid surface phase may undergo some alteration in chemical composition. Depending on the structure of the catalyst this might be noticeable as a small change in mass which can sometimes be observed utilizing thermogravimetry as already noted. The process of heterogeneous catalysis is thus first adsorption at adjacent positions on the surface of reactant molecules with sufficient energy to form the adsorbed activated complex. There is a balance here of forces, for if the reactants are too strongly adsorbed then the other reactant or reactants will not be able to have access to the surface and the reaction is retarded. The process is essentially kinetic in that the two (or more) reactants must be adsorbed adjacent to each other on the surface to form the activated complex which can then be desorbed as a different species. Such surface sites may form only a small fraction of the total available surface. The number of these sites may be enhanced by the addition of promoters or blocked off and reduced by the presence of catalytic poisons.

The role of TG in investigating such systems may be explored a little further. In the solid-solid-gas systems (where the catalyst is a solid, and either one of the reactant or product phases is a solid) then TG can be used to follow the actual course of the reaction. The production of oxygen by the decomposition of potassium chlorate using the catalyst manganese dioxide is an example. A possible reaction sequence illustrating the role played by the manganese dioxide was given by MacLeod[3] as

$$2MnO_2 + 2KClO_3 \rightarrow 2KMnO_4 + Cl_2 + O_2$$
$$2KMnO_4 \rightarrow K_2MnO_4 + MnO_2 + O_2$$
$$K_2MnO_4 + Cl_2 \rightarrow 2KCl + MnO_2 + O_2$$

This theory explains why traces of chlorine are found in the product gas. Other alternative routes for this decomposition may be found with other catalysts which are just as effective.[4,5]

However where the catalyzed reaction involves gaseous reactants with a solid catalyst phase, TG may only be used to follow the reaction if the product involves a solid phase. Properly designed balances enabling vacuum and controlled ambient pressures to be maintained allow accompanying processes of chemisorption and physical adsorption to be studied. It is then possible to tackle the problem in such cases by studying either the rate of adsorption or the equilibrium condition, i.e. the adsorption isotherm.

2 THERMODYNAMIC FEATURES OF CATALYSTS

A consideration of the basic thermodynamic concepts regarding catalysis would seem to be in order before going on to describe the application of thermal analysis to the topic. This and the following section on kinetic factors will address the above points in terms of whether they are adequate in present usage.

It has already been stated that a catalyst (considered here to be an accelerator of a process) can only affect a reaction which would take place anyway. For a reaction to be possible the process must be accompanied by a decrease of free energy. Thus the function of a catalyst is to make the overall process take place by an alternative path involving a smaller free energy of activation. If a reaction occurs in several stages, then under the influence of a catalyst the free energy of activation of the slowest of these stages must be appreciably less then for the reaction in the absence of a catalyst.

It is of course possible for the Gibbs free energy to be negative not just for a single reaction but for several reactions involving the same chemical species. This explains why different catalysts and different experimental conditions can produce different products. Thus a silver catalyst is capable of acting as a catalyst on ethylene to produce (depending on the experimental conditions):[6]

1. ethylene oxide (if chemisorbed oxygen was present)

2. acetaldehyde

or 3. carbon dioxide and water.

Ethyl alcohol can be converted, using different catalysis and experimental conditions to:

1. C_2H_4 using Al_2O_3 at 300°C

2. CH_3CHO using Cu or Ni at 300-400°C

or 3. $C_2H_5-O-C_2H_5$ using Al_2O_3 at 257-300°C

In all these processes the product depends on the catalyst but in the temperature range used the Gibbs free energy is negative. Thus for a particular reaction

$$-\Delta G° = RT\ln Kp$$

where $\Delta G°$ is the Standard Gibbs Free Energy and Kp the equilibrium constant (written here as for a gaseous reaction) with R the Gas Constant and T the temperature (in Kelvin) of the process. The variation of $\Delta G°$ with temperature can be stated in terms of the reaction isochore

$$\frac{d(\ln Kp)}{dT} = \frac{\Delta H°}{RT^2}$$

where $\Delta H°$ is the standard enthalpy change at constant pressure. The corresponding entropy change (ΔS) is given by

$$\Delta G = \Delta H - T\Delta S$$

The equilibrium constant (Kp) is equal to the quotient of the forward and reverse reactions and can be written as

$$Kp = \frac{kf}{kr}$$

where kf is the forward reaction rate constant and kr that for the reverse reaction. It follows that a catalyst accelerates equally the rates of the forward and the reverse reactions in any simple chemical transformation.

It follows from these basic premises that the effectiveness of a catalyst in any particular reaction can be judged by the use of either DTA, DSC or TG in simply deciding the temperature at which the reaction is observed for any series of catalysts. The most effective catalyst is simply that which promotes the reaction at the lowest temperature. In DTA we have an effective way of investigating heterogeneous gas catalyzed reactions simply by putting the catalyst in the sample cell and passing the mixture of gases involved through the unit under a controlled temperature regime.

3 KINETIC FACTORS

There are many detailed publications available on this topic.[7-14] The following steps may be involved in heterogeneous catalysis and any one of these steps may be rate-determing. The steps are:

(i)   transport of reactants to catalyst surface

(ii) adsorption of reactants on the catalyst, e.g.

A(g) + catalyst(s) ⇌ catalyst - A(ads)

B(g) + catalyst(s) ⇌ catalyst - B(ads)

(iii) interaction of adsorbed species

A(ads) + B(ads) ⇌ X(ads) + Y(ads)

[A(ads), B(ads), X(ads) and Y(ads) indicate adsorbed reactant and product species.]

(iv) desorption of product from catalyst surface

Catalyst - X(ads) ⇌ Catalyst + X

Catalyst - Y(ads) ⇌ Catalyst + Y

(v) transport of the products away from the surface.

Steps (i) and (v) involve no chemical reaction but depending on the system being considered, especially with reference to the experimental conditions, can be rate-limiting. If any of the other steps are rate-determining then the process might be expected to obey an Arrhenius type reaction

$$k = A \exp (-E/RT)$$

where k is the specific reaction rate constant, A is the pre-exponential factor and E is the Activation Energy. By classical theory this latter term is a measure of the amount of energy which reactant molecules possess before they overcome the energy barrier between them and the product state. The pre-exponential term in gas-phase reactions is conventionally supposed to specify the fraction of molecules having kinetic energy greater than E. In dealing with an order-type reaction, the rate of reaction (r) at time (t) (under isothermal conditions) is given by:

$$r = kP_A{}^M P_B{}^N$$

where P_A and P_B are the pressures of reactants A and B at time (t), and M and N denote the order of reaction with respect to A and B. The term k is thus the rate when the reactants are at unit pressure. The transition state theory explains the general process by supposing that a transition state exists designated in the general reaction process as:

$$A + B \underset{k2}{\overset{k1}{\rightleftharpoons}} AB^{\ddagger} \overset{k3}{\rightarrow} X + Y$$

Here A and B are reactants, X and Y the products and constants k1, k2, k3 the specific reaction rate constants for the indicated processes. This transition state is assumed to be in equilibrium with the reactants even though the overall chemical reaction is irreversible, so that

$$\frac{k1}{k2} = kp^{\ddagger} = \frac{P_{AB}{}^{\ddagger}}{P_A P_B}$$

The rate of reaction is given by the rate of decomposition of the transition state complex, thus

$$r_f = k_3 P_{AB}{}^{\ddagger}$$

In heterogeneous catalysis the catalyst surface provides an alternative site where the transition state complex occurs providing a route of lower activation energy. This complex can be considered to decompose when a vibrational mode changes to a translational mode since this represents an irreversible vibration. The frequency of such a change is

$$\vartheta = \frac{e}{h}$$

where h is Planck's constant and e is the average energy of the relevant vibration. The frequency of vibration ϑ of the disintegrating bond can be calculated using the Planck relationship

$$e = h\vartheta = \frac{RT}{N}$$

The value of e can be assessed as the bond dissociation energy. This is a simplified description; the detailed and proper procedure is given by Laidler.[15] The end result is the same and ϑ is identified with k_3, so that

$$r_f = \frac{RT}{Nh} k_p^{\ddagger} P_A P_B$$

The formal equation takes the form

$$r_f = kP_A P_B$$

when $k = \frac{RT}{Nh} k_p^{\ddagger}$

In thermodynamic parameters we have:

$$-\Delta G^{\ddagger} = RT\ln k_p^{\ddagger} \quad \text{and} \quad \Delta G^{\ddagger} = \Delta H^{\ddagger} - T\Delta S^{\ddagger}$$

where $\Delta G^{\ddagger}$ is the free energy of activation, $\Delta H^{\ddagger}$ is the standard enthalpy of activation, and $\Delta S^{\ddagger}$ is the standard entropy of activation. Thus

$$k = \frac{RT}{Nh} \exp \frac{-\Delta G^{\ddagger}}{RT} = \frac{RT}{Nh} \exp \frac{\Delta S^{\ddagger}}{R} \exp \frac{-\Delta H^{\ddagger}}{RT}$$

Comparison with the Arrhenius equation leads to the following identities

$$-\Delta H^{\ddagger} \equiv E \quad \text{and} \quad \frac{RT}{Nh} \exp \frac{\Delta S^{\ddagger}}{R} \equiv A$$

In a catalyzed reaction the specific reaction rate must be increased, with a decrease in the free energy of activation of the reaction ($\Delta G^{\ddagger}$) given by

$$k = \frac{RT}{Nh} \exp \frac{-\Delta G^{\ddagger}}{RT}$$

caused by alterations in the value of the entropy ($\Delta S^{\ddagger}$) and the enthalpy ($\Delta H^{\ddagger}$). The value of $\Delta S^{\ddagger}$ associated with the activated state in the catalyzed reaction will be less than that associated with the activated state in the uncatalysed reaction. The reason for this is that in the catalyzed reaction the transition complex will be sited on a catalyst surface, when there will be a consequent loss of translational freedom. For good catalytic action the $\Delta H^{\ddagger}$ value must be decreased by an amount greater than the change in entropy. The temperature dependence of the rate for the corresponding homogeneous and heterogeneous reactions plotted using the Arrhenius equation are shown in Figure 1. It is then apparent that both A and E are lower in the heterogeneous catalyzed reaction.

4 STOICHIOMETRY OF A CATALYST DURING USE

The usual definition of a catalyst includes a statement that it is unchanged chemically but this may not be true in heterogeneous catalysis. This is because the exchange process at the surface may involve surface structural atoms or ionic species. The rate of replacement of surface species may not match the rate at which these species are utilized during catalysis. This would be in disagreement with the usual definition of the catalyst.

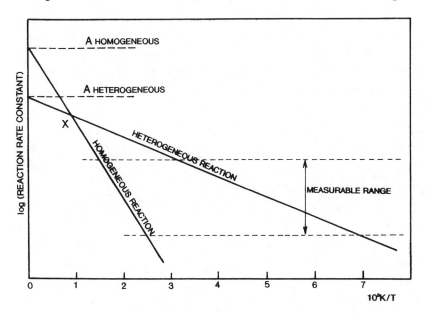

Figure 1 Arrhenius Plots for Heterogeneous and corresponding Homogeneous Reaction
Note: Point X – in many similar reactions with different catalyst there is a tendency for all lines to cross at a single point x to give a compensation effect.

However the more basic concept that the catalyst allows a chemical reaction to take place via a more energetically favorable reaction path is preserved.

Most heterogeneous catalysts are metals, oxides or sulfided oxides. One prerequisite of such systems is the preparation of the catalyst with a large surface area. Metals with large surface areas are very reactive so such catalysts must be protected or supplied as oxides which have to be reduced in the catalytic reactor. Sulfided catalysts are sulfided *in situ* with the oxides as a precursor.[16] It is not clear why the sulfide catalysts have to be prepared in this way except the pragmatic reason that they are better catalysts than the "massive" sulfide preparations.

There are a variety of ways in which thermal analysis can be used to characterize the catalyst, to study the preparative methods involved and the enthalpy changes occurring in the preparative methods and the actual catalyzed reaction. In the case of oxide and sulfide systems the use of TG shows the non-stoichiometric nature of the prepared catalyst and manner in which the stoichiometry varies with the experimental environment. In metal systems as catalysts thin layers of oxide may passivate the catalyst, or the surface phase may differ from the composition and structure of the bulk. This is especially true for alloys.[17] The composition of the surface may vary from one surface position to another.

The small changes in mass occurring in the catalyst during the reaction may be followed using the TG technique. In oxide catalyst systems the catalysis process occurs by a process of oxygen exchange. The oxygen may be present as a chemisorbed layer or incorporated in the lattice structure from which it may be abstracted during the catalyzed process. The oxygen used in the catalyst process must be replaced, usually from atmospheric oxygen. Over long periods of usage the process of using of the oxygen available from the catalyst almost always exceeds the rate of replacement with a consequent decrease in catalyst activity. Regeneration of the catalyst is by a separate process usually by reheating in oxygen when gasification of carbonaceous residues and other surface inhibitors occurs followed by reincorporation of oxygen in the catalyst lattice.[18] The use of TG to study such systems is obvious.

5 METAL CATALYSTS

In the case of metal catalysts the catalysis step is often the addition or removal of a molecule of hydrogen. Many metal surfaces can dissociate molecular hydrogen into hydrogen atoms.

There are two modes of catalytic adsorption on metal surfaces - dissociative and associative. Dissociative chemisorption occurs with hydrogen:

$$H_2 + 2M_{(cat)} \rightarrow 2H.M_{(cat)}$$

There are various alternatives:

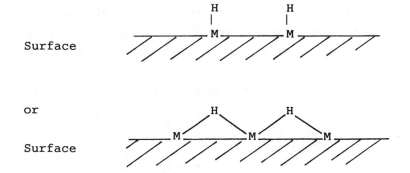

Associative adsorption occurs when the adsorbed species is not disrupted:

e.g.

$$
\begin{array}{ccc}
\quad H & \quad H & \quad H\ H \\
\quad | & \quad | & \quad |\ | \\
R'-C = & C-R''+2M \rightarrow & R'-C-C-R'' \\
 & & |\ | \\
 & & -M-M- \\
 & & |\ |
\end{array}
$$

If the bonding forces in chemisorption are too strong then desorption of the products of catalysis can not take place. The generalization is that for catalysis to take place the adsorption bond must be strong enough to provide adequate coverage but not so strong that subsequent formation of a transition complex and the desorption of products is rendered impossible. This generalization is not confined to metal catalysis but applies to all catalytic processes.

In this field TG is a valuable tool in following the reduction of oxides *in situ* to the metal catalyst. A typical DSC method would be to use a pressure DSC cell and to study the chemisorption processes by the exotherm produced in the temperature region from 75°C to around 250°C. This gives the heat of chemisorption which needs correction for the heat of physical adsorption. This is particularly important when the metals are deposited on substrates. In the Pd-Pt catalysts this type of investigation showed that loss of activity by the secondary metal can synergistically affect the total catalyst activity.[19,20]

As chemisorption is confined to a monolayer then with some knowledge of the stoichiometry of the surface reaction it may be possible to determine surface areas. Thus hydrogen and carbon monoxide adsorption have been used to determine the surface areas of metal catalysts.[21]

6 OXIDE CATALYSTS

Oxide catalysts can be divided into three groups; n-type semi-conductor oxides, p-type semi-conductors and those that are regarded as stoichiometric oxides. The semi-conductor oxides are characterized by a degree of non-stoichiometry. In these oxide systems the stoichiometry varies during the catalyst process and TG experiments (as already noted) can be used to follow the behavior of the solid phase. TG experiments can also be used to follow the stoichiometry of such oxides when heated in various atmospheres. If such data are combined with measurements of electrical conductivity of the sample during the heat treatment this information can be very helpful. The change in conductivity is to be noted, not the actual electrical conductivity of the material. It is not sufficient to say that n-type semi-conductor oxides lose oxygen on heating whilst p-type semi conductor oxides gain oxygen on heating as the experimental conditions should be specified. Experiments of the above nature using normal TG equipment provide semi-conductor characterization but may also give an indication of the catalytic mechanism. In oxide catalysis where oxygen is involved at reactant, transition or product stage the oxide species can be attached to a surface metal either as $O_{(ads)}$ or O^{2-}. In the latter case this may appear only in the adsorbed surface layer or incorporated via diffusion into the lattice structure. Thermal desorption experiments may reveal that the oxide incorporates both species desorbed in two easily recognisable temperature regions. These comments may be considered by reference to two catalyzed reactions[22,23] namely:

$$2N_2O \rightarrow 2N_2 + O_2 \quad \text{and} \quad 2CO + O_2 \rightarrow 2CO_2$$

Nickel oxide is a catalyst in the first reaction and its action may be represented as

$$N_2O + Ni^{2+}_{(surface)} \rightarrow N_2 + O^-_{(ads)} \cdots Ni^{3+}_{(surface)}$$

This can be compared with the chemisorption of oxygen on such surfaces

$$0.5O_2 + Ni^{2+}_{(surface)} \rightarrow O^-_{(ads)} \cdots Ni^{3+}_{(surface)}$$

The activity of p-type oxides in the decomposition of N_2O is in the same order as their ability to adsorb oxygen. Two mechanisms of desorption are available

$$2(O^-_{(ads)} \cdots Ni^{3+}_{(surface)}) \rightarrow O_2 + 2Ni^{2+} \quad \text{or}$$

$$N_2O + O^-_{(ads)} \cdots Ni^{3+} \rightarrow O_2 + N_2 + Ni^{2+}$$

The above reaction pathways are not available for

insulator and n-type oxides which are much less active.

Two reaction pathways are available for the oxidation of carbon monoxide in the presence of oxygen.[24,25] The use of NiO serves as an example of a p-type oxide catalyst, where oxygen is first chemisorbed

$$0.5O_2 + Ni^{2+}_{(surface)} \rightarrow O^-_{(ads)} \cdots \cdots Ni^{3+}_{(surface)}$$

followed by reaction with adsorbed carbon monoxide

$$O^-_{(ads)} \cdots \cdot Ni^{3+}_{(surface)} + CO_{(ads)} \rightarrow CO_2 + Ni^{2+}$$

Zinc oxide serves as an example of n-type activity in this reaction leading to low temperature catalysis, via the process

$$CO + 2O^{2-}_{(surface)} \rightarrow CO_3^{2-}_{(surface)} + 2e$$

with a consequent reduction of Zn^{2+}

$$Zn^{2+}_{(surface)} + 2e \rightarrow Zn_{(surface)}$$

The decomposition of the $CO_3^{2-}_{(surface)}$ leads to regeneration of one of the two oxide ions involved

$$CO_3^{2-}_{(surface)} \rightarrow CO_2 + O^{2-}_{(surface)}$$

The $Zn_{(surface)}$ species can be regenerated by oxygen to Zn^{2+}

$$0.5O_2 + Zn_{(surface)} \rightarrow O^{2-}_{(surface)} + Zn^{2+}_{(surface)}$$

A vacuum microbalance can be used to investigate the unbalance between the rate of regeneration of the $O^{2-}_{(surface)}$ species and the rate at which it is used up to produce the product. The regeneration procedure can be followed using a vacuum microbalance in a TG mode at controlled pressures.

The manganese-copper double oxides, and other double oxides show a catalyst selectivity which is a reflection of the availability of the lattice oxide ions.[26] Compound oxides such as bismuth molybdate show a similar selectivity.[27] The stoichiometric oxides such as aluminosilicates have a catalytic action which depends variously on their affinities for water, their acidic character or their molecular sieve properties.[28]

7 SUPPORT MATERIALS

Support materials have the primary use that if the catalyst material is spread over them as a thin film then the surface area of the catalyst is enhanced. The basic requirements are then that the support material should have a large available area, be relatively inert chemically and prepared in such a manner that it has a reasonable mechanical strength. Typical catalyst precursors are inorganic solids such as nitrates,[29] hydroxides,[30] carbonates,[31] etc. These are converted to the oxide catalyst by thermal decomposition which can be studied by thermal analysis. In certain circumstances, with oxalates or formates in nitrogen the metal may be the end product of heat treatment.[32] Usually however in order to produce supported metal catalysts the common industrial usage is to produce the oxide first and then produce the metal catalyst by reduction *in situ* in a catalyst reactor. Supported sulfide catalysts are also produced by sulfiding processes carried out *in situ* on an oxide precursor. The thermal decomposition route provides for many oxysalts, reactant oxides with large surface area as measured by adsorption of vapors and gases.[33] The surface area developed depends not only on the oxysalt being decomposed but also on the method of thermal treatment.[34] The production and control of solid state activity is discussed in several books.[35-38] The kinetics of these thermal decompositions has been reported in great detail by Brown, et al.[39]

Two alternative processes are available to produce the coating of the catalyst precursor on the support surface. The first involves precipitation from solution onto an inert support. The second requires the inert support to be soaked in a soluble salt solution of the metal species required as the catalyst.[40] The salt normally chosen is the nitrate. The insulator type oxides are typical support materials such as silicas, and aluminas. Using the nitrate solution the soaked material is dried to leave the nitrate on the support material surface. This is followed by more severe heat treatment to decompose the nitrate to leave an oxide deposited on and attached to the inert surface. A combination of surface area determinations and thermal analysis provides an excellent method of studying these systems. Two examples of inert support materials are discussed here but there are other inert supports forming important industrial catalysts.

8   ALUMINA

Alumina is a typical support material.[41] Aluminum hydroxide ($Al(OH)_3$) or $Al_2O_3.3H_2O$) occurs in three common forms, gibbsite (γ-$Al(OH)_3$), bayerite (α-$Al(OH)_3$) and nordstrandite. Two forms of the hydrate, $Al_2O_3.H_2O$ are also described, boehmite (γ-AlOOH) and diaspore (α-AlOOH). The use of DTA to show the effect of heat treatment on these hydrate forms has been reviewed by Mackenzie and Berggren.[42]

The aluminum hydroxide is produced as a gelatinous precipitate by the action of ammonia or other alkali on aluminum sulfate or chloride solutions. This is represented by the reaction equation:

$$Al_2(SO_4)_3 + 6NH_4OH \rightarrow 2Al(OH)_3 + 3(NH_4)_2SO_4$$

The resultant gel has a great excess of water above that indicated by the formula. A freshly precipitated gel will dissolve in acids forming aluminum salts but an aged gel dissolves with difficulty. The "gel" dissolves in caustic soda but not in ammonia to form the aluminate ($NaAlO_2$). Active aluminas are the high surface area materials prepared by heat treatment of the gel.[43] Heating to 800°C or above leads eventually to the formation of corundum (α-Al_2O_3) with a very small surface area. The neutralization of sodium aluminate solution with carbon dioxide forms the basis of preparing gibbsite and bayerite.[44-46] In a diverse TG investigation of precipitates of hydrated aluminum oxides obtained from various reaction schemes a wide

variety of behavior patterns is observed.[47] Some typical
TG curves are given in Figure 2 and Table 1 indicates
the temperature at which dehydration has proceeded to
Al_2O_3.

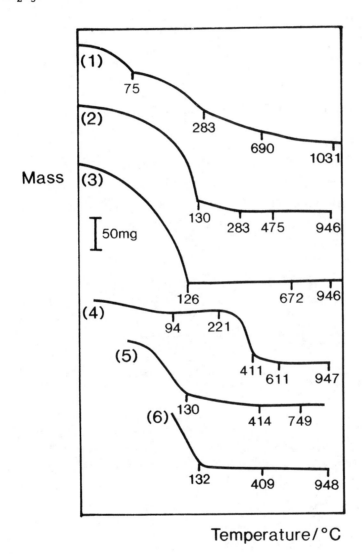

Figure 2 Typical TG curves for drying Alumina
Precipitates. Precipitating Reagent : 1. Aqueous
Ammonia, 2. Gaseous Ammonia, 3. Urea, 4. Urea/Succinic
Acid, 5. Ammonium Sulfide, 6. Ammonium Carbonate
Taken from ref. 47

Table 1   Summary of TG curves of Hydrated Aluminum Oxide
          Precipitates

Precipitation Method	Type of Precipitate	Min. Temp. ($^\circ$C) at which Al_2O_3 formed
Aqueous ammonia	gelatinous	1031
Gaseous ammonia	easy filtration	475
Urea	good	672
Urea/succinic acid	good	611
Mercury chloroamide	unsuitable method	676
Hexamine	good filtration	473
Pyridine	good filtration	478
Ammonium acetate	good	475
Ammonium formate		539
Ammonium succinate	$Al(OH)_3$ can be filtered at 100°	509
Ammonium carbonate	good	409
Ammonium bicarbonate	good	514
Hydrazinium carbonate	compact easily filtrable $Al(OH)_3$ at 122°C	524
Carbon dioxide	rapid filtration	945

Note in this last preparation continued CO_2 passage
can result in a carbonated gel structure used in
pharmaceutical preparations.

The Table indicates the variability of the precipitate
quality. It is a good example of the effect of the
experimental conditions of precipitation on the
characteristic features of material. Thus material
produced by precipitation by bromine below 60°C had a
drying temperature as low as 280°C.

The heat treatment of these precipitates can result
in two groups of active aluminas as follows:
(i) low temperature aluminas (γ-group): $Al_2O_3 \cdot nH_2O$,
where n is given as $0 < n < 0.6$ obtained by dehydrating at
temperatures up to 600°C

(ii) high temperature aluminas (δ-group): nearly anhydrous Al_2O_3 obtained by heating between 900 and 1000°C.

In summary the TG technique should show details of the dehydrations, the transformations in the oxide should be detectable by high temperature DTA but in most cases X-ray diffraction methods are used.

9 SILICA

Silica used as a catalyst support may generally be regarded as polycondensation products of orthosilicic acid. Various forms of silica exist which can be recognized by DTA/DSC techniques. Three forms exist; tridymite, cristobalite and quartz. The phase change between these forms involves reasonably large movements of atoms in the solid lattice and a considerable kinetic factor is involved in such transitions. DTA is not a good technique to measure these primary transitions. However there are more minor changes, "inversions", which involve only small alterations in structure, in each of these three forms, which produce a good "signal" on the DTA trace and can be used quantitatively and qualitatively to estimate the amount of each phase present. The work of Fenner[48] is quoted in Table 2. The active silicas with high surface areas are often quoted as being amorphous. It might seem that DTA could be used to identify the phase in such circumstances, and although this has been claimed, the endothermic loss of water and possible changes from regions of disorder to regions of order make these identifications difficult.

There are various terms which refer to these amorphous and porous silicas. Silica gel generally refers to the preperation where the sample has been formed in a liquid medium but this is generally water. The pore structure might become filled with alcohol in some preparations when the term alcogel can be applied. Aquagel or hydrogel refers to a gel with the pores filled with water. A xerogel is a gel from which the

<u>Table 2</u> Transitions between quartz (Q), tridymite (T), and Cristobalite (C)

Transition Temperature/°C	Crystalline Change
117	$\alpha T \leftrightarrow \beta_1 T$
163	$\beta_1 T \leftrightarrow \beta_2 T$
198-241	$\beta C \to \alpha C$
220-275	$\alpha C \to \beta C$
570	$\beta Q \to \alpha Q$
575	$\alpha Q \to \beta Q$
870±10	$Q \leftrightarrow T$
1470±10	$T \leftrightarrow C$
1625	$\beta C \leftrightarrow$ silica glass

α and β indicate high and low temperature forms respectively

liquid has been removed from the pores but with some damage and collapse to the porous structure. The term amorphous silica covers all these forms and is used here.

There are three general methods of preparation:

1. precipitation from silicate solutions (generally by acid)[49-52]

2. hydrolysis of silicon derivatives such as silicon tetrachloride[53]

3. pyrogenic silica preparations (called aerosils). This involves vapor hydrolysis of silicon halides or related techniques.[54]

The water associated with the silica may be incorporated into the structure especially at the surface and also simply fill up the porous structure. The first product of the condensation of silicic acid is a hydrogel. The subsequent drying process produces the xerogel which is the usual industrial form of the porous silica. The preparation, stability and adsorption properties of porous silica have been reviewed by Okkerse,[55] and the overall chemistry by Iler.[56]

The gel or precipitate as collected contains water

held between particles, water in various sized pores present in the structure and "combined" water (probably present as hydroxyls). This complexity presents difficulty in interpreting TG data. The properties associated with the surface are determined by the character, number and distribution of the silanol groups present. There are methods involving TG, the adsorption of polar molecules, the reactions of various compounds with the hydroxyl groups and IR spectroscopy studies which all lead to an understanding of the arrangement of hydroxyl groups at the surface.

The adsorbed water can be physically adsorbed and chemisorbed. As already noted the chemisorbed or bound water refers to OH groups bound in various ways to the silicon atoms. Surface structures of this kind are

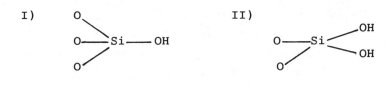

I)

II)

and III)

It is probable that II and III condense to SiOH groups on "normal" drying. According to de Boer and Vleeskens[57] a silica dried at 120°C under atmospheric conditions has lost all physisorbed water and still contains all surface hydroxyls. Above this temperature there is progressive loss of water until around 1000°C all the water from these hydroxyls is lost. If the surface area is known (from nitrogen adsorption experiments) then the hydroxyl groups per square millimicron (N_{OH}) can be calculated. Iler[56] gives the formula for this calculation as

$$N_{OH} = \frac{660 \ W}{S}$$

where S is the surface area in m^2g^{-1} and W is % mass of water lost between 120° and 1000°C. There is conflict of opinion that 120° represents the temperature which distinguishes between loss of physically bound water and the loss of hydroxyls. A further complication is that beyond some temperature around 650°C the surface area might be modified by sintering. Iler suggests a theoretical value of 8 OH nm^{-2}. However lower values are commonly reported and a commonly observed value for rehydrated silicas at high temperatures, then fully rehydrated and dried is 4.6 OH nm^{-2}.

10   COMMENT

The above descriptions and topics are illustrative of some important aspects of the application of thermal analysis in studying catalysts and chemisorption. However there are many other aspects of catalysis where thermal analysis finds use but space precludes a discussion in depth. Thus in the above two sections two support systems were discussed but others are important industrially.   Important among those is magnesia.[58,59] Carbon is another material; this is of importance in catalysis for a variety of reasons, as an adsorbent, as a deposit on catalysts causing deactivation, and as a support material.[60-62]

Autocatalysis in thermal decomposition is another subject where thermal analysis can be applied with success. Certain oxalates show this type of behavior.[63] Zinc oxalate dihydrate can be quoted as an example

$$ZnC_2O_4 \cdot 2H_2O \xrightarrow[\text{Endothermic}]{\text{Air and } N_2} ZnC_2O_4 + 2H_2O$$

$$ZnC_2O_4 \xrightarrow[\text{Exothermic}]{\text{Air}} ZnO + CO + CO_2$$

and

$$\text{ZnC}_2\text{O}_4 \xrightarrow[\text{Endothermic}]{\text{Nitrogen}} \text{ZnO} + \text{CO} + \text{CO}_2$$

The exothermic reaction in air is due to catalysis caused by the appearance of ZnO. This is a good catalyst for the reaction:

$$2\text{CO} + \text{O}_2 \xrightarrow[\text{catalyst}]{\text{ZnO}} 2\text{CO}_2$$

This reaction is exothermic, and the overall process is thus also exothermic.

The effect of additions, i.e. dopants, to the decomposition of various oxysalts is shown to produce catalyst effects by studying both TG and DTA data, and if possible interpreting the data in terms of kinetic analysis based on non-isothermal techniques.[64] Examples are the decomposition of potassium permanganate[65] and manganese (II) carbonate.[66]

Poisoning of catalysis is another area where thermal analysis provides good data. Water is a common source of poisoning in catalyst systems. It is for example essential to eliminate water in the oxidation of carbon monoxide to carbon dioxide over manganese oxides in the presence of air.[26]

It is certain that in these areas there is a great scope for an expansion of thermal analysis applied to the subject of catalysis, catalysts and chemisorption.

11 REFERENCES

1. S. Glasstone, 'The Elements of Physical Chemistry', Macmillan, London, 1955, p.620.
2. D. Dollimore, A.K. Galwey, and G. Rickett, <u>J.Chim Phys.</u>, 1975, <u>72</u>, 1059.
3. G. McLeod, <u>J. Chem. Soc.</u>, 1889, <u>55</u>, 184; 1894, <u>65</u>, 202.
4. J.S. Booth, D. Dollimore and G.R. Heal, <u>Thermochim. Acta</u>, 1980, <u>39</u>, 281, 293.
5. V.V. Boldyrev, Z.G. Vinokurova, L.N. Senchenko, G.P. Shchetinina and B.G. Erenburg, <u>Russ. J. Inorg.Chem.</u>, 1970, <u>15</u>, 1341.
6. G.H. Twigg, <u>Proc. Roy. Soc., A</u>,1946, <u>188</u>, 92, 105, 123.
7. G.C. Bond, <u>Discuss. Farad. Soc.</u>, 1966, <u>41</u>, 200.

8. G.C. Bond, 'Catalysis by Metals', Academic Press, London, 1962.
9. G.C. Bond, 'Principles of Catalysis', Monograph for Teachers, No.7, The Royal Society of Chemistry, 1972.
10. G.C. Bond, 'Heterogeneous Catalysis Principles and Applications', 2nd Edition, Clarendon Press, Oxford, 1987.
11. P.G. Ashmore, 'Catalysis and Inhibition of Chemical Reactions', Butterworths, London, 1903.
12. C.N. Satterfield and T.R. Sherwood, 'The Role of Diffusion in Catalysis', Adderson-Wesley, Reading, 1963.
13. J.M. Thomas and W.H. Thomas, 'Introduction to the Principles of Heterogeneous Catalysis', Academic Press, London, 1967.
14. E.K. Rideal, 'Concepts in Catalysis', Academic Press, London, 1968.
15. K.J. Laidler, 'Chemical Kinetics', McGraw-Hill, New York, 2nd edn, 1972, p.72.
16. P.C.H. Mitchell in C. Kemball (Ed.), 'Catalysis', Vol.1, The Chemical Society, London, 1977, p.37.
17. R.L. Moss in C. Kemball (Ed.) 'Catalysis', Vol. 1, The Chemical Society, London, 1977, p.37.
18. J.R.H. Ross and J.M. Thomas (Eds.), 'Surface and Defect Properties of Solids', Vol. 4, The Chemical Society, London, 1975, p.34.
19. W.E. Collins, <u>Calorimetry</u>, 1970, <u>2</u>, 353.
20. G.C. Bond and D. Webster, <u>Ann. NY Acad. Sci.</u>, 1969, <u>158</u>, 541.
21. H. Charcosset, C. Bolivar, R. Frety, R. Gomez and Y. Trambouze, 'Proc. Vac. Microbal Tech.', (eds. S.C. Bevan, S.J. Gregg and N.D. Parkyns) Vol.2, Heyden, 1973, 175.
22. E.C. Pitzer and J.C.W. Frazer, <u>J. Phys. Chem.</u>, 1941, <u>45</u>, 761.
23. E.R.S. Winters, <u>J. Catal.</u>, 1969, <u>15</u>, 144.
24. G.M. Schwap and J. Block, <u>Z. Phys. Chem.</u>, 1954, <u>1</u>, 42.
25. G. Parravano and M. Boudart, 'Advances in Catalysis', Vol.7, Academic Press, New York, 1955, p.50.
26. D. Dollimore and K.H. Tonge, <u>J. Chem. Soc. A</u>, <u>1970</u>, 1728.
27. R. Higgins and P. Heyden, in C. Kemball (Editor), 'Catalysis', Vol. 1, The Chemical Society, London, 1977, p.168.
28. M.S. Spencer and T.V. Whittam in C. Kemball and D.A. Dowden (Eds.), 'Catalysis', Vol.3, The Chemical Society, 1980, p.189.
29. D. Dollimore, G.A. Gamlen and T.J. Taylor, <u>Thermochim. Acta</u>, 1985, <u>86</u>, 119.
30. K.A. Broadbent, J. Dollimore and D. Dollimore, <u>Thermochim.Acta</u>, 1988, <u>33</u>, 131.
31. A.S. Bhatti, D. Dollimore and N. Blackmore, <u>Thermochim. Acta</u>, 1984, <u>79</u>, 205.

32 D. Dollimore, Thermochim. Acta, 1991, <u>177</u>, 59.
33. D. Dollimore and D. Nicholson, J. Chem. Soc., <u>1962</u>, 960.
34. D. Dollimore, Thermochim. Acta, 1989, <u>148</u>, 63.
35. A.K. Galwey, 'Chemistry of Solids', Science Paperbacks, London, 1967.
36. S.J. Gregg, 'The Surface Chemistry of Solids', Chapman and Hall, London, 1961.
37. V.V. Boldyrev, M. Bulens and B. Delmon, 'The Control of the Reactivity of Solids', Elsevier, Amsterdam, 1979.
38. R. Sh. Mikhail and E. Robens, 'Microstructure and Thermal Analysis of Solid Surfaces', Wiley, Chichester, 1983.
39. M.E. Brown, D. Dollimore and A.K. Galwey, 'Reactions in the Solid State', Vol. 22 of Comprehensive Chemical Kinetics (C.H. Bamford and C.F.H. Tipper, eds.), Elsevier, Amsterdam, 1980.
40. D. Dollimore, in 'Proc. of the Workshop on the State of the Art of Thermal Analysis', Gaithersburg, MD, May 1979, Spec. Publ. Natl, Bur. Stand, 1980, p.1.
41. B.G. Lippens and J.J. Steggerda, in 'Physical and Chemical Aspects of Adsorbents and Catalysts', (Editor B.G. Linsen), Academic Press, London, 1970, Ch.4, p.171.
42. R.C. Mackenzie and G. Berggren in 'Differential Thermal Analysis', Vol.1, (Editor R.C. MacKenzie), Academic Press, London, 1970, p.479.
43. H.C. Stumpf, A.S. Russell, J.W. Newsome and C.M. Tucker, Ind. Eng. Chem., 1950, <u>42</u>, 398.
44. L.E. Oomes, J.H. de Boer and B.C. Lippens, 'Reactivity of Solids', (Editor J.H. de Boer), Elsevier, Amsterdam, 1961, p.317.
45. H. Ginsberg, W. Huttig, and H. Stiehl, Z. Anorg. Allg. Chem., 1961, <u>309</u>, 233; 1962, <u>318</u>, 238.
46. T. Sato, J. Appl. Chem. (London), 1962, <u>12</u>, 553.
47. C.L.Duval, 'Inorganic Thermogravimetric Analysis', Elsevier, Amsterdam, 1953, p.105.
48. C.N. Fenner, Am. J. Sci., 1913, <u>36</u>, 331.
49. D. Dollimore and G.R. Heal, Trans. Farad. Soc., 1963, <u>59</u>, 2386.
50. D. Dollimore and T. Shingles, J. Colloid InterfaceSci., 1969, <u>29</u>, 601.
51. D. Dollimore and T. Shingles, J. Appl. Chem., 1969, <u>19</u>, 218.
52. D. Dollimore and T. Shingles, J. Chem. Soc. A, 1971, 872.
53. F.E. Bartell and Y. Fu, J. Phys. Chem., 1929, <u>33</u>, 276.
54. A.G. Amelin, Kolloidn Zh., 1967, <u>29</u>, 16.
55. C. Okkerse in 'Physical and Chemical Aspects of Adsorbents and Catalysts', (Editor B.G. Linsen), Academic Press, London, 1970, Ch.5, p.213.
56. R.K. Iler, 'The Chemistry of Silica', 1979,
57. J.H. de Boer and J.M. Vleeskens, Koninkl. Ned. Akad.Wet. Proc., 1957, <u>B60</u>, 23, 45, 54; 1958, <u>B61</u>, 2, 85.

58. W.F.N.M. De Vleesschauser in 'Physical and Chemical Aspects of Adsorbents and Catalysis', (Editor B.G. Linsen), Academic Press, London, 1970, p.265.
59. D. Dollimore and P. Spooner, Trans. Farad Soc., 1971, 67, 2750.
60. Th. Van der Plas in 'Physical and Chemical Aspects of Adsorbents and Catalysts', (Editor B.G. Linsen), Academic Press, London, 1970, p.425.
61. R.L. Bond (Editor), 'Porous Carbon Solids', Academic Press, London, 1967.
62. A.N. Ainscough, D. Dollimore and G.R. Heal, Carbon, 1973, II, 189.
63. D. Dollimore, Thermochim. Acta, 1987, 117, 331.
64. D. Dollimore, T. Evans, Y.F. Lee and F.W. Wilburn, 'Proc. 19th North Amer. Thermal Anal. Soc.', (Editor I.R. Harrison), 1990, p.373.
65. J.S. Booth, D. Dollimore and G.R. Heal, Thermochim.Acta, 1980, 39, 293.
66. D. Dollimore and P.R. Rogers, Thermochim. Acta, 1979, 30, 273.

Quality Assurance in the Thermal Analysis Laboratory

P. H. Willcocks

ICI MATERIALS, WILTON MATERIALS RESEARCH CENTRE, WILTON,
MIDDLESBROUGH, CLEVELAND TS6 8JE, UK

1 INTRODUCTION

Quality Assurance (QA) is fundamental to the commercial,
technical, or even legal requirements of the industrial
world. Thermal Analysis (TA) is merely one of the
technique areas now becoming increasingly used to support
one, or all of these QA factors. In order to achieve these
criteria the calibration, testing and analysis procedures
must comply with strictly controlled standards and hence
there is a need to incorporate Quality Assurance within
the technique area and within all the TA Laboratories
involved.

2 QUALITY ASSURANCE IN THERMAL TECHNIQUES

General Considerations

Thermal Analysis techniques have become increasingly used
over recent years. This is without doubt a feature which
reflects the advances in 'Quality' programs undertaken by
companies worldwide. Suppliers and manufacturers have been
able to take advantage of the relatively low budget TA
instrumentation available to evaluate their materials and
methods of production. There is a potential to examine all
the stages within a process, from raw materials through to
the final products.

Quality initiatives have been prevalent in all the
European countries and during the next few years, with the
onset of the single market in 1992, there will almost
certainly be an increase in the need for common attitudes

and approaches to Quality Assurance. Companies, large or
small, which fail to recognise the need to operate within
accredited schemes such as BS5750[1], ISO9000[2], etc. may in
the longer term find their competitiveness difficult to
maintain. Thermal Analysis will be part of these Quality
processes.

One significant outcome apparent from quality programs carried out so far is the need to obtain 'equivalent
data' from wide ranges of different laboratories which are
nominally involved in the same techniques. Use of quality
systems such as NAMAS[3], GLP[4], and BS5750 are designed to
remove many of the intrinsic problems, but do not within
themselves provide an absolute guarantee of success. The
standardisation of the laboratories and their methods of
working are perhaps the key to it all. These aspects alone
are already being investigated at various levels eg. the
Community Bureau of Reference (BCR)[5] , Brussels is active
in investigating the standardisation of laboratories within the European Community. Such concepts are not of course
unique to Thermal Analysis and the principles can be
applied to the vast majority of laboratory testing worldwide[6]. The aim of this paper is to bring together two
increasingly inter-related themes which are 'the use of
Thermal Analysis to provide Quality Assurance' and 'techniques of Quality Assurance applied to Thermal Analysis'.
The latter is the foundation on which 'equivalent data'
might be achieved.

Five of the most common thermal techniques will be
considered. These are Differential Scanning Calorimetry
(DSC), Differential Thermal Analysis (DTA), Thermomechanical Analysis (TMA), Thermogravimetry (TG) and Dynamic Mechanical Analysis (DMA). Examples of some QA applications will be be presented and others can be found elsewhere in this publication. Discussion of what problems can
be associated with each of these measurements will be put
forward, particularly where such problems have been, or
are potentially, of major importance because of the
relatively wide areas of application. Even though the

examples will be limited to a few applications, the
general thoughts and principles are likely to be
applicable to most of the Thermal Analysis related methods
currently in use.

Differential Scanning Calorimetry

Differential Scanning Calorimetry (DSC) is perhaps one of
the most commonly used of all the thermal techniques for
QA purposes. Industrial users include manufacturers,
suppliers and purchasers of Pharmaceuticals, Plastics, Ex-
plosives, Inorganic and Organic Chemicals, ... The main
use of DSC will vary for each industry, but, from simple
determination of a melting range, through measurement of
enthalpy related data, to complex kinetics, the same critical variation in energy or temperature differentials as
functions of temperature and time are the fundamental
criteria on which the result will depend.

Within the plastics industry, the largest single user
of DSC instrumentation, the ability to monitor how semi-
crystalline or amorphous polymeric components change, or
are likely to change, at a given temperature, can be vital
information. Consider then, a commonly used DSC technique
in which a semi-crystalline polymer is cooled at a fully
controlled rate, from the melt, ie. from a state where all
previous thermal history is known to have been removed. In
any one laboratory the temperatures recorded for the start
and the peak temperature of the exotherm observed for the
crystallisation process are frequently used to characterise the material. This may be to monitor a change in the
average molecular weight, the effect of residues or nucleating agents, or the degree of shear received during
processing. For results obtained using a given instrument,
one on which the control samples have been assessed, the
technique has been shown to be excellent in a QA role. It
can be situated close to the manufacturing plant to enable
the evaluation of batches of material as they are produced
and rapidly determine whether or not each batch meets the
specification established for a good quality product. This
type of method can be incorporated into part of a wider

quality program (BS5750, ISO9000) for that production
area. It essentially relies on the fact that the given
instrument will be capable of reproducing any of the
temperatures recorded for the crystallisation exotherm to
say ±1°C. This precision can only be achieved using very
carefully managed preparation and analysis procedures.
For the polymer, an overall tolerance of say ±3°C could
then become the QA limit for the test, an acceptable level
of variability to avoid problems in end use applications.
But, the question which arises is, what might happen if an
instrument is changed? To many users a DSC instrument set-
up accurately according to the manufacturers specification
will provide very precise, accurate and reproducible temperature data. Since temperature has been accurately calibrated, using standard reference materials from a traceable source, the output data is generally assumed to be
consistent. Unfortunately, this is not always seen to be
true. Experience has shown that, whilst the temperatures
recorded during the heating scans can be relatively easily
monitored and checked against the values quoted for the
standard reference materials, cooling presents the analyst
with many different problems. In the first instance there
is no known, reliable, 'cooling reference standard' ie. a
pure material with well defined thermal behaviour observed
during controlled cooling that can be accurately reproduced to fractions of a degree Centigrade. Assessment of
what is the 'real temperature' of a specimen during cooling is therefore generally the responsibility of the thermal analyst seeking high quality data. Many analysts have
been aware of this problem for a long time and each has
attempted to provide solutions within a laboratory; the
stage has now been reached where these individual ideas
should become part of a collaborative, inter-laboratory
justification.

Cynics will ask what the significance of these temperature differences might be to something like the plant
QA method previously described. This will obviously be
dependent on the likely error recorded using different

instrumentation. The author can report observed temperature differences of almost 10°C in the relative positions
of the exotherms for the same materials examined using
different instrumentation. Most alarming, however, is the
fact that the instrumentation on which the widest temperature differentials were recorded were nominally the same
model, calibrated as recommended. The materials would have
been examined using identical methods. Specimen size,
shape and preparation, purge gas and basic calibration
procedures would have been very tightly controlled. Hence,
in reply to the original question regarding the effect of
instrument changes in the QA method, it can be seen that
any large discrepancy in recorded temperatures would
result in rejection of an otherwise good material, or
acceptance of a poor sample, by the QA section. To the
industrial user this may have important downstream costs
implications and loss of credibility in the marketplace.

What are the reasons for an apparent 10°C discrepancy
in temperatures when the instrument precision is quoted to
something like 0.1°C? The outcome has very definite implications, not only for QA within the thermal laboratory,
but also has some relevance to all scientific testing
methods involving the use of fully computerised analytical
instrumentation. The error was found to have resulted from
a combination of 'firmware' and 'software' modifications
which can of course be made without reference to, or the
knowledge of, the end user. The differences that occur are
not necessarily obvious in the standard setting-up procedures recommended in the manufacturers' manuals and are
often discovered only by chance. It emphasises the point
that, whilst computerisation is essential to every scientist, it still becomes necessary to know how the input
data might control and relate to the eventual output data.
For the thermal analyst it is essential to know what system is being used to generate the temperature values which
are fundamental to all thermal measurements; some difficulties exist with regard to the latter even when in a
nominal 'isothermal' mode of operation.

This example was intended to illustrate that a simple procedure, performed routinely in hundreds of different laboratories each working day may not be as absolute as might have been expected. Hopefully, it will be a first stage in making the thermal analyst conscious of what limitations could exist in a QA method and also why the results might vary in inter-laboratory testing. There is also a requirement that the instrument manufacturers fully understand the operation of their machines such that, should a problem arise, at the very least, an explanation for the effect is available.

Comments were made earlier that results generated using a single instrument should always be consistent providing that the machine has been regularly calibrated and maintained. Whilst this is largely true, experience has also shown that the constant thermal cycling can effectively 'age' some of the hardware components in any TA equipment. This has the effect of changing the characteristics of say the measuring cell with time/usage. Calibration checks on a very regular basis, part of any QA procedure, will ensure some degree of reproducibility. It should be remembered that many of the firmware/software parameters, including those for linearisation of the temperature response of the cell, are established for an 'unaged' system. With time these parameters may need modifying but, in reality, are rarely changed. Again there can be an influence on output which directly affects the consistency of results.

Other Thermal Techniques

Temperature cannot be as precisely measured, relative to the specimen, for TG, TMA or DMA instrumentation. These techniques rely on sensor or control thermocouples/resistance thermometers remote from the specimen and, in general terms, have much greater furnace volumes which mean poorer temperature distribution. For the QA role the temperature precision is not considered to be a major problem since the requirement may be to monitor temperature differences to $\pm2\,^{\circ}C$ at best. In something like a TMA for example it is sometimes only the relative temperature difference which is measured and not necessarily the absolute value at any given instant. Where improved resolution of the temperature is required the sensors must be mounted as close as is practically possible to the specimen being evaluated. It is one of the key design features of any 'constant rate' thermal analytical devices which were discussed elsewhere in this publication.

Dynamic Mechanical Analysis (DMA) is widely used in a Quality Assurance role. Examples of where it might be used are to examine the state of cure of a cross-linkable resin system, the rate of cure of a resin as a function of temperature, or Modulus vs Temperature profiles for components to be used in the Aerospace industry. It can be used as a performance check in the Heat Distortion testing of mouldings since it too is measuring the effective stiffness of a material as a function of temperature. As was discussed earlier, in the vast majority of these tests the specimens would be relatively large, in heat transfer terms, and the furnaces, by necessity, would also be large. Hence, any QA testing would involve very slow heating rates, say less than 2°C/min. As in all thermal testing the large specimen size, combined with low thermal conductivity or diffusivity properties, tend to reduce the achievable measurement precision. It effectively makes the comparative testing used in QA work very dependent on the actual specimen dimensions and its position within the instrument furnace. The nature and range of the DMA equipment makes it difficult to use traceable standard reference materials in order to calibrate the modules. Temperature monitoring is often best done using certified or accredited thermocouples, independently mounted and ideally, for calibration purposes, embedded in different positions within a typical specimen. As with all mechanical testing, especially at elevated temperatures, the clamping of the specimen is a key factor in the determination of modulus; hence, clamp pressure, clamp condition (flatness) and size must all be considered. As the temperature is altered the

compressive modulus of the specimen is likely to change
with regard to temperature or time. This will affect the
degree of clamping and will have a direct influence on the
accuracy of any measured moduli. Many of the makers of
dynamic mechanical equipment, including equipment which is
primarily used to measure rheological changes, are
currently working towards improved calibration techniques
and reference materials. The request for improvements in
the precision and accuracy within this area came mainly
from those involved in establishing QA tests as part of
quality accreditation requirements.

Thermomechanical Analysis (TMA) is another technique
which, although widely used in QA type applications, is
often difficult to calibrate fully in every aspect of the
range of operation. Temperatures can be fairly precisely
determined using traceable reference materials since
specimen size can be small. However, the furnaces do not
generally have good temperature distribution characteristics and specimen position relative to the thermocouple
is again critical. The force on the specimen and the measured displacements are also difficult to quantify with
precision. The estimate of the accuracy/precision that
the thermal analyst believes can be achieved is the
important factor; these may not match the quoted specifications.

Atmosphere Effects

Virtually all the QA work involving thermal analysis
techniques will be carried out using a 'purge gas' to
provide a controlled atmosphere in terms of chemistry and
often heat transfer considerations. The condition and flow
rate of these gases (eg. argon, nitrogen, helium, oxygen)
are obviously an important consideration. It affects any
chemical change occurring within the specimen, or in
thermogravimetry can alter the 'buoyancy' effect on the
TG balance. The buoyancy effects result from changes in
the gas density (static buoyancy) and aerodynamic drag
with temperature and are dependent on gas type/amount, the
size of the crucible and specimen, heating rates etc.

The purity of 'cylinder gases' is generally accepted to be within a controlled specification, but locally piped main supplies may contain higher levels of moisture or hydrocarbons derived from the compression systems. The contaminated nature of the latter may have detrimental effects on the analysis in any of the thermal systems, but particularly in TG, DSC or DTA type evaluations. Care must be taken with the filtration or cleaning of purge gas supplies since many of the systems supplied for this purpose can introduce traces of other compounds into the gas. Also, any form of valve switching can produce a pressure pulse which, for very sensitive TA equipment, can be detected via the output signal. Electrical interference is sometimes blamed for poor quality thermal scans when in fact it is the variability in the purge gas supply which is responsible.

One aspect of purge gas supply with direct influence on a QA technique was recorded when examining the degradation of a polymer at a temperature typical of that used in extrusion or injection moulding. It was known that the polymer became cross-linked when degraded and that the degree of the cross-linking reaction could be monitored using a relative change in the recrystallisation and remelting parameters. A control sample, which would normally remain stable and therefore unchanged for residence times >60mins, was seen to cross-link in less than 10 minutes. The effect was unexpected and in fact could not be reproduced on another DSC system. The latter used the same purge gas supply, hence this was not thought to be a factor. However, after many other tests indicated that only a contaminated purge supply could be responsible it was found that the DSC used for the initial tests had an inherent design fault which allowed small amounts of air to be drawn into the DSC cell. This only occurred at critical flow rates of the purge gas and therefore was a difficult problem to eliminate totally without redesigning the DSC cell itself.

Hermetically sealed systems are sometimes used to

avoid the effects of oxidation during scans. What is often
not done is to ensure that the encapsulation/sealing
process is actually done in an inert atmosphere prior to
testing. Even though the amount of trapped oxygen might be
considered irrelevant it has been shown to be significant.
Analysis of an organic solvent demonstrated this very
effectively. During initial heating in a DSC a small, yet
reproducible exotherm was recorded at a given temperature.
To increase the resolution the specimen size was doubled;
the result was then to reduce the size of the process. Why
had this happened? It was subsequently established that
residual oxygen, from air sealed along with the solvent,
was responsible. The solvent reacted, exothermically, with
the oxygen during a heating scan. Increasing the size of
the specimen reduced the effective headspace within the
crucible above the solvent and hence, the amount of oxygen
available for the reaction. The direct result of the
latter would be to reduce the size of the exotherm.

   3   FUNDAMENTAL ISSUES

The use of Thermal Analysis for Quality Assurance is well
established. Some issues relevant to the techniques have
been presented, mainly to raise the awareness to some of
the problems which can, and often do, exist. In many
instances actions are currently being undertaken, or are
being considered, to overcome such problems, hopefully on
a global scale. To the new or inexperienced user of
thermal testing some of the points made may have raised
doubts as to the validity of this type of testing in a QA
role. The fact remains that there are massive benefits to
be gained but, in line with all scientific testing, the
limitations or pitfalls in any given technique must be
taken into account. Consultation with those people who
are experienced in this field of application, the use of
a collaborative approach and the correct standardisation
of calibration, operation and evaluation procedures are
now seen to be the key issues.

   Operation under accredited schemes such as BS5750,

ISO9000, NAMAS, etc., will ensure a systematic approach to analysis, the traceability of standard reference materials, procedures and data plus the guarantee that calibration is consistently maintained. Compliance with these actions should result in the 'equivalent data' sought within the European Community for collaborative laboratory testing. To a small company, the costs of registration/accreditation may be restrictive, but well documented procedures which ensure a clear, well directed and fully traceable system will enable reliable data to be produced. Fundamental to much of this is the requirement to employ staff with not only the right degree of academic ability, but also with the correct attitude and level of commitment, ie. the ability to recognise exactly what is needed to produce reliable QA data, why this is important and the motivation to achieve Quality performance. As always there is a small price to pay and perhaps within the TA laboratory this will be the additional time required to support the increased bureaucracy of a well managed system.

What of the future? Any list of requirements or objectives is likely to be endless. In setting up 'quality systems' within a Thermal Analysis laboratory problems arise continuously as each new procedure is investigated. Established methods of calibration, the frequency of performance checks, even the accuracy of the balance used are the sort of factors which suddenly have to be con-sidered in far greater depth. The following are merely a few pointers to what might become significant over the next few years in Thermal Analysis environments where Quality Assurance is a prime requirement.
a) Well defined, documented Standard Reference Materials particularly in areas such as TG, TMA, DMA. These are essential to establish the precision in temperature, displacement, moduli etc. and form the basis of all calibration and performance check procedures.
b) Increased use of collaborative or inter-laboratory testing, particularly within, and between, the larger

international companies and the Standards committees.

c) Validation of software for control and data processing
particularly when this has been produced under licence by
an independent software house remote from the instrument
supplier. Firmware may also eventually be scrutinised.

d) Increased standardisation of the terminology, procedures, precision assessment etc. of all aspects related
to thermal measurement. Ideally, this would become the
responsibility of a central controlling body, to be
comprised of representatives from various industrial and
academic institutions.

e) Improved training of staff in both the operation and
understanding of the techniques and methods associated
with TA.

f) Regular auditing and reviews of the working systems
within laboratories. This can be achieved internally,
or as part of wider quality systems such as NAMAS, GLP
and similar accredited schemes.

The achievement of fully justified, quality results
must be the aim of all those involved in Quality Assurance
testing and is essential to the concept of a quality
product. A first stage in this process is to raise the
awareness to what is likely to be required. Hopefully, the
latter has been encouraged by some of the thoughts and
ideas discussed, with particular relevance to the many
aspects of Thermal Analysis.

REFERENCES

1. BS5750 is a product of the British Standards
    Institution and is published in various parts as
    follows.
    BS5750 Part 1 : 1987 - Quality systems, specification
    for design/development, production, installation and
    servicing. (Equivalent to ISO9001 - 1987 and EN29002
    - 1987).

BS5750 (Quality Systems) Part 2 : Specification for production and installation. (Equivalent to ISO9002 - 1987 and EN29002 - 1987).
BS5750 (Quality Systems) Part 3 : Specification for final inspection and test. (Equivalent to ISO9003 - 1987 and EN29003 - 1987).
BS5750 (Quality Systems) Part 4 - 1990 : Guide to BS5750 (formerly BS5750 parts 4,5, & 6).

2. ISO9000 - 1987 : Quality Management and Quality Assurance Standards - Guidelines for Selection and Use. This standard is issued by the International Organisation for Standardisation. It clarifies the quality concepts and provides guidelines for the selection and use of International Standards, either for internal quality management purposes (ISO9004) or for external quality purposes (ISO9001, 9002, 9003).

3. NAMAS - National Measurement Accreditation Service is operated by the National Physical Laboratory. The main basic documents are the NAMAS regulations, document M11 and NAMAS accreditation criteria, document M10. These are used in conjunction other more specialised pamphlets issued by the NAMAS Executive. The aim is to provide accreditation within laboratories for specific testing or calibration procedures to ensure that high quality work is carried out, that agreed or specified methods have been followed and that all measurements are traceable to national and international standards. (Other countries are introducing equivalent quality organisations).

4. GLP - Good Laboratory Practice. It can be a legal requirement that data obtained to meet certain national regulations is measured in accordance with GLP principles. The latter are established by various national authorities (eg. FDA, DOH) and are based in the main on an OECD document 'Good Laboratory Practice in the Testing of Chemicals', published 1982. The UK version is set-out in 'Good Laboratory Practice, The United Kingdom Compliance Program', published by the Department of Health in 1989. These principles have been supplemented by advisory leaflets on the application of GLP to computer systems, field studies and the role of quality assurance. Laboratories are inspected by the DOH.
Mutual acceptance agreements exist between UK and USA, UK and Japan.

5. BCR - Community Bureau of Reference (DG XII), Rue de la Loi, 200, B-1049 Brussels. This was established by the European Commission to help laboratories in member States provide accurate, reliable and equivalent methods, procedures and data. It is involved in inter-laboratory test programmes and provides Standard Reference Materials.

6. 'A Journey Through Quality Control', Broderick, Cofino, et al. Mikrochim.Acta [Wien] 1991, II, 523-542

The International Confederation for Thermal Analysis—A Review

S. St. J. Warne

DEPARTMENT OF GEOLOGY, THE UNIVERSITY OF NEWCASTLE, CALLAGHAN,
NSW 2308, AUSTRALIA

1 INTRODUCTION

The International Confederation for Thermal Analysis (ICTA) is a well established international scientific organization whose statutes delineate its aims to be to promote international understanding and co-operation in the science of thermal analysis.

To date this has been achieved in several main ways.

(1) The holding, at regular intervals, currently every 4 years, of international scientific congresses. The papers at these are published as formal proceedings of each congress which facilitates the dissemination and widespread availability of these current contributions in the field of thermal analysis. Such congresses have been held in Aberdeen, Scotland (1965), Worcester, USA (1968), Davos, Switzerland (1971), Budapest, Hungary (1974), Kyoto, Japan (1977), Bayreuth, Germany (1980), Kingston, Canada (1982), Bratislava, Czechoslovakia (1985) and Jerusalem, Israel (1988), while the 10th congress will be held at Hatfield, UK in 1992.

(2) The work of the Scientific Commission which is composed of a number of scientific committees whose work in specific areas of thermal analysis is vital for its advancement, promulgation and the vitally important area of educational preparation for thermal analysts.

(3) The publication, (bi monthly) of Thermal Analysis Reviews and Abstracts (TAR and A) and ICTA News (twice a year) and, when appropriate, individual publications such as "For Better Thermal Analysis".

(4) The encouragement of national thermal analysis societies which when affiliated with ICTA are represented by one member on its council. In addition there are 4 members of council termed "Councillors-at-Large". These represent parts of the world not as yet represented by an affiliated society. This category of council members has been implemented to make available as widely as possible the knowledge and participation which ICTA brings to its members.

A schematic indication of the structure by which ICTA operates is set out in Figure 1, which appeared originally in the ICTA publication "For Better Thermal Analysis III".[1]

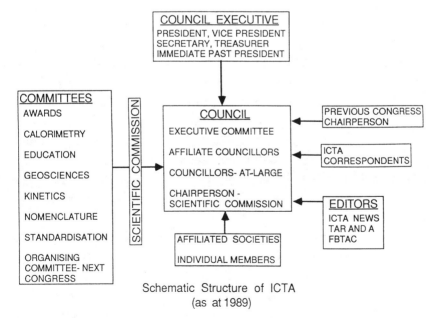

Schematic Structure of ICTA
(as at 1989)

<u>Figure 1</u>   Schematic Structure of the International Confederation for Thermal Analysis after J.O. Hill, Ed. 1991.[1]

2 MEMBERSHIP

Membership of ICTA is open to all who have an interest in thermal analysis and calorimetry, individuals, companies and affiliated societies, on payment of a modest annual fee to the Membership Secretary from whom further details may be obtained i.e. Dr H.G. McAdie, McAdie and Associates, Suite 340, 245 Eglinton Ave East, Toronto, Ontario M4P 3B7, Canada.

Membership benefits include involvement with leading thermal analysts and calorimetrists in 38 countries and from 19 affiliated national societies, a list of the names and addresses of all members, which is invaluable for scientific consultation, collaboration and visits. The receipt of ICTA news and a copy of "For Better Thermal Analysis", reduced registration fees for ICTA Congresses together with eligibility to nominate or be nominated for ICTA Awards.

3 AWARDS

These are awarded at the four yearly ICTA Congresses.

The TA Instruments Award consists of a plaque, an honorarium of 1000 US dollars and expenses to the Congress. Recipients who are considered to have made outstanding contributions to the science are expected to present a lecture to the Congress.

The Young Scientists Award consists of a plaque, financial support to attend the Congress, where they present their award winning paper. Candidates must be under 35 year of age and submit a paper on their own research results.

The Chairman of the ICTA Awards Committee is currently Dr D.J. Morgan, British Geological Survey, Keyworth, Nottingham BH12 5GG, UK.

Honorary Membership of ICTA is restricted to a maximum of 6 in total. It is awarded for sustained and outstanding contributions to ICTA and nominations may come from the affiliated societies or 6 currently financial ICTA members.

4 THE SCIENTIFIC COMMISSION (See Figure 1)

This is currently headed by Dr J. Rouquerol (France) and is composed of the following committees which are followed by the names of the chairpersons. Awards (Dr D.J. Morgan UK), Calorimetry (Prof G. Della Gatta, Italy), Education (Prof Edith A Turi, USA), Geosciences (Prof W. Smykatz-Kloss, Germany), Kinetics (Dr J.H. Flynn, USA), Nomenclature (Dr C.J. Keattch, UK), Standardization (Prof E.L. Charsley, UK), Organizing Committee for the next ICTA Congress, i.e. Hatfield UK August 1991 (Prof D. Nowell, Department of Chemistry, Hatfield Polytechnic, Hatfield, Hertfordshire, AL10 9AB, UK).

The following are brief summaries of the basic activities of these committees, the prime source of which is the publication "For Better Thermal Analysis III",[1] to which reference should be made for further details.

The Awards Committee composed of a total of 5 members whose responsibility it is to prepare shortlists of candidates to be submitted to ICTA Council for final selection of award winners.

It is worth mentioning, in this context, that although not under the auspice of ICTA a number of other awards at the national society level are made and have contributed greatly to the encouragement and recognition of thermal analysts specifically and thermal analysis in general i.e. the Bodenheimer Award (Israel), the Czechoslovakia Working Group for Thermal Analysis Award (Czechoslovakia), the Mettler-NATAS Award (USA), the Netsch-GEFTA Award (Germany), the Netsch-ITAS Award (India), the PL Thermal Sciences-ITAS Award (India), the Maple-TAWN Award (Netherlands), and the Thermal Methods Group Award (UK).

The role of the Calorimetry Committee is to promote all aspects of calorimetry, particularly in collaboration with thermal analysis research, and the establishment of a uniform framework of reporting, experimentation and research with the inclusion of calorimetry wherever suitable within the scope of ICTA.

The relatively new Education Committee is one typified by much energy, expansion, achievement and successes in truly pioneering work with the aid of an enthusiastic international network of subcommittees dedicated to the establishment and growth of high quality thermal analysis courses at all levels upon which the future vigour, acceptance and application scope of thermal analysis rests.

The even newer Geosciences Committee is in its first term of operation having been set up at the 9th ICTA Congress in Jerusalem in 1988. In response to its aims of publicising the role of thermal analysis in earth science investigations and research and between thermal analysts and geoscientists, it has employed a different approach, viz. by using the process of specific topic meetings of experts who are to make contributions in their particular area of thermal analysis expertise in the geosciences. The first of these was held in Karlsruhe at the beginning of October 1990 and the resultant initial volume,[2] "Thermal Analysis in the Geosciences" was published less than a year later, while the next volume has already been initiated.

The Kinetics Committee is involved with the development of suitable nomenclature, symbols, practices and terms for the classification and description of processes, the organization of Kinetics Symposia, to facilitate and improve the quality of kinetics for thermal analysts and compile kinetic data and applications for practical purposes, for example, endurance testing or hazard analysis.

The brief of the Nomenclature Committee is to continuously review the definitions, terminology and methods within the field of thermal analysis for use internationally so that reporting, description and results are of a uniform and readily understandable nature.

The Standardization Committee has, to date, succeeded

in establishing a number of "Standard Reference Materials" for the calibration of many thermal analysis parameters and methods. This work is ongoing in response to increasing ranges of operation and the development of new methods. It is achieved through a network of internationally based laboratories which facilitate the testing and evaluation of proposed standard reference materials.

The ICTA Council (see Figure 1) is thus composed of the Executive Committee, Affiliated Councillors, Councillors-at-Large and the Chairperson of the Scientific Commission. There remains only the composition of the Executive Council i.e. the President, Prof S.St.J. Warne (Australia), Vice President, Dr T. Ozawa (Japan), the immediate Past President, Prof H.J. Seifert (Germany), the Secretary, Prof S. Yariv (Israel) and the Treasurer, Prof P.K. Gallagher (USA).

From this brief review it may be seen that ICTA is actively involved in a large number of aspects of thermal analysis through its truly international membership.

5 REFERENCES

1. J.O. Hill, (Ed.), 'For Better Thermal Analysis and Calorimetry III', International Confederation for Thermal Analysis UK, 1991.

2. W. Smykatz-Kloss and S.St.J. Warne (Eds.), 'Thermal Analysis in the Geosciences', Springer Verlag, Heidelberg, 1991.

Sources of Information in Thermal Analysis

E. L. Charsley

THERMAL ANALYSIS CONSULTANCY SERVICE, LEEDS METROPOLITAN UNIVERSITY,
CALVERLEY STREET, LEEDS LS1 3HE, UK

1 INTRODUCTION

The purpose of this chapter is to provide a brief introduction to sources of information in thermal analysis which it is hoped will be of value to newcomers to the field. For a more comprehensive overview, the handbook *For Better Thermal Analysis and Calorimetry*, Ed. J.O.Hill, 3rd Edition, International Confederation for Thermal Analysis, 1991, may be consulted.

2 THERMAL ANALYSIS LITERATURE

Books

The shortage of modern thermal analysis books has been addressed in recent years and an interesting feature has been the emergence of several self-instruction texts. For a general introduction to thermal analysis techniques the book by Brown is recommended, while for a more comprehensive treatment the text by Wendlandt may be consulted. Although out of print, the two volumes edited by Mackenzie are widely available in technical libraries and provide a wealth of information on the applications of differential thermal analysis, particularly in the fields of inorganics and minerals.

General Texts

Introduction to Thermal Analysis, M.E.Brown, Chapman &

Hall, 1988, 211pp.

Thermal Analysis, 3rd Edition, W.Wm.Wendlandt, Wiley, 1986, 814pp.

Differential Thermal Analysis, Ed. R.C.Mackenzie, Academic Press, Vol.1, 1970, 775pp., Vol.2 1972, 607pp

Self-instruction Texts

The Practice of Thermal Analysis, G.van der Plaats, Mettler, 1991, 101pp.

Thermal Analysis, B.Wunderlich, Academic Press, 1990, 450pp.

Thermal Methods, Analytical Chemistry by Open Learning, J.W.Dodd & K.H.Tonge, Wiley, 1987, 337pp.

Specialist Texts

DSC on Polymeric Materials, Netzsch Annual for Science and Industry, Vol. 1, E.Kaisersberger & H.Möhler, Selb, 1991, 84pp.

Thermal Analysis in the Geosciences, Eds. W.Smykatz-Kloss & S.St.J.Warne, Springer Verlag, 1991, 379pp.

Pharmaceutical Thermal Analysis, J.L.Ford & P.Timmins, Ellis Harwood, 1989, 313pp.

Thermal Characterisation of Polymeric Materials, Ed. E.I.Turi, Academic Press, 1981, 972pp.

ASTM Special Technical Publications. The American Society for Testing and Materials (ASTM) has published a number of Special Technical Publications (STP) based on symposia on specific areas of thermal analysis, sponsored by ASTM Committee E-37 on Thermal Measurements:-

Purity Determinations by Thermal Methods, ASTM STP 838, Ed. R.L.Blaine & C.K.Schoff, 1984, 151pp.

Compositional Analysis by Thermogravimetry, ASTM STP 997, Ed. C.M.Earnest, 1988, 293pp.

Materials Characterization by Thermomechanical Analysis, ASTM STP 1136, Ed. A.T.Riga & C.M.Neag, ASTM, Philadelphia, 1991, 195pp.

These books are available from ASTM, 1916 Race Street, Philadelphia PA 19103, U.S.A. or ASTM European Office, 27/29 Knowl Piece, Wilbury Way, Hitchin, Herts SG4 OSX.

Reviews

In addition to reviews in the thermal analysis and specialist journals, a very useful general review of thermal analysis is published in the Fundamental Review issue of Analytical Chemistry which appears every two years. The most recent review is by D.Dollimore, Thermal Analysis, Analytical Chemistry, Fundamental Reviews, 62, 1990, 44R.

Journals

There are two journals entirely devoted to thermal analysis and calorimetry:-
Journal of Thermal Analysis, Editors: Dr.J.Simon & Dr.B.Androsits, Technical University, 1521 Budapest, Hungary.
Publishers: John Wiley & Sons Ltd., Baffins Lane, Chichester, Sussex PO19 1UD & Akadémiai Kiadó, PO Box 245, H-1519 Budapest, Hungary.
Thermochimica Acta, Editor: Prof.W.W.Wendlandt, Dept. of Chemistry, University of Houston, Houston, Texas 77004, U.S.A. Publishers: Elsevier Science Publishers B.V., P.O.Box 211, 1000 AE Amsterdam, The Netherlands.

In addition to original research papers and short communications, both journals are increasingly being used to publish the proceedings of thermal analysis conferences.

Technical Literature from Thermal Analysis Instrument Manufacturers

The majority of thermal analysis instrument manufacturers offer a comprehensive range of technical information sheets and application notes which provide a valuable source of thermal analysis data. These are normally available free of charge. A list of manufacturers is given in section 7.

3 ABSTRACT SERVICES

The major source of abstracted thermal analysis data is *Chemical Abstracts* published by the American Chemical Society who also produce *CA selects: Thermal Analysis*. The latter is a current awareness service specifically on thermal analysis topics and is published fortnightly by Chemical Abstracts Service, 2540 Olentangy River Road, P.O.Box 3012, Columbus, Ohio, 43210, U.S.A. (the UK representatives are the Royal Society of Chemistry, Thomas Graham House, Science Park, Milton Road, Cambridge CB4 4WF.)

  Thermal Analysis Reviews & Abstracts (TARandA) is an official publication of ICTA and is published 6 times a year by Interscience Communications Ltd, 24 Quentin Road, London, SE13 5DF (before this the abstracts were published as *Thermal Analysis Reviews* by Wiley) Publication will cease at the end of 1991. Although, unlike Chemical Abstracts, TARandA cannot be computer searched, the keyword section relating to technique enables searches to be carried out on specific areas of interest to thermal analysts such as atmosphere, heating rate and furnace design.

4. THERMAL ANALYSIS SOCIETIES

The international body for thermal analysis is the *International Confederation for Thermal Analysis* (ICTA) whose activities are reviewed in a separate chapter in this book. Details of membership may be obtained from the ICTA Membership Secretary, Dr. H.G.McAdie, H.G. McAdie Associates, Suite 340, 245 Eglington Avenue East, Toronto, Ontario, M4P 3B7, Canada.

Some 17 National Societies are affiliated to ICTA and provide a forum for scientific activities and meeting at a national level. In the U.K. the Society is the *Thermal*

Methods Group of the Analytical Division of the Royal Society of Chemistry, (Secretary Dr.C.J.Keattch, P.O. Box 9, Lyme Regis, Dorset DT7 3BT). Names and addresses of other National Societies may be obtained from the ICTA Membership Secretary at the above address.

5   INTERNATIONAL CONFERENCES

ICTA holds an international congress every four years and in 1992 this will be in the U.K. in Hatfield, Herts. The proceedings will be published in a special edition of the Journal of Thermal Analysis.

Alternating with the ICTA Congress is the *European Symposium on Thermal Analysis and Calorimetry* (ESTAC). ESTAC 6 will be held in September 1994 in Grado near Trieste, Italy. The organising Chairman is Prof. A.Cesaro, Dipartimento di Biochemica Biofisica e Chimica Della Macromolecole, Universita Degli Studi di Trieste, Piazzale Europa 1, I-34127 Trieste, Italy.

The proceedings of earlier conferences in both these series form a valuable source of TA literature. A list is given in the book by Brown and in *For Better Thermal Analysis and Calorimetry*.

6   STANDARDS AND STANDARD METHODS

With the involvement of many laboratories in quality assurance programmes and the general recognition of the care needed to obtain high quality thermal analysis data, there has been an increasing interest in calibration materials and standard methods for thermal analysis measurements.

Standard Reference Materials

International Confederation for Thermal Analysis. The Standardisation Committee of ICTA has developed a series of Certified Reference Materials (CRM's) for temperature calibration of DTA/DSC equipment covering the range -89°C to 925°C. A set of magnetic CRM's are available for the temperature calibration of thermobalances. The CRM's are marketed by the National Institute of Standards and Technology, Office of Standard Reference Materials, Room 205, Building 202, National Institute of Standards and Technology, Gaithersburg, MD 20899, U.S.A. They may also be obtained from the Laboratory of the Government Chemist (see below).

Laboratory of the Government Chemist. The Laboratory of the Government Chemist offers a series of organic materials, with certified enthalpies of fusion, covering the temperature range 81°C to 147°C, for the calibration of DSC equipment. The enthalpies of fusion of the materials were determined by adiabatic calorimetry and their temperatures of fusion are also certified. They can be obtained from the Office of Reference Materials, Laboratory of the Government Chemist, Queens Road, Teddington, Middlesex, TW11 0LY, U.K.

National Institute of Standards and Technology. The National Institute of Standards and Technology in addition to marketing the ICTA CRM's offers a series of Standard Reference Materials for temperature and enthalpy calibration of DSC instruments. Three of the materials - mercury, tin and zinc - have been measured by precision calorimetry.

A set of materials is also available for the evaluation of methods of purity determination by DSC.

Standard test procedures

A useful set of Standard Methods has been developed by Committee E37 on Thermal Measurements of the ASTM. These standards are listed in *Vol.14.02, 1991 Annual Book of ASTM Standards*, ASTM, 1916 Race Street, Philadelphia, PA 19103, U.S.A. A number of other ASTM Committees also have interests in thermal analysis. These include:- D-2 (Petroleum Products and Lubricants), D-5 (Coal and Coke), D-9 (Electrical Insulating Materials), D-20 (Plastics) and E-27 (Hazard Potential of Chemicals).

7 THERMAL ANALYSIS EQUIPMENT MANUFACTURERS

The following is a list of a number of companies supplying thermal analysis equipment in the U.K. The name of the parent company is given first, followed by the U.K. representative in brackets.

Cahn Instruments Inc., 16207 S.Carmenita Road, Cerritos, CA 90701, U.S.A. (Scientific and Medical Products Ltd., Shirley Institute, 856 Wilmslow Road, Didsbury, Manchester M20 8RX)

Linkam Scientific Instruments Ltd, 8 Epsom Downs Metro Centre, Waterfield, Tadworth, Surrey KT20 5HT, U.K.

Mettler-Toledo AG, CH-8606, Greifensee, Switzerland (Mettler-Toledo Ltd., 64 Boston Road, Beaumont Leys Leicester LE4 1AW)

Netzsch Geratebau GmbH, Wittelsbacherstrasse 42, P.O.Box 1460, D-8672 Selb/Bavaria, Germany.(Netzsch Mastermix Ltd., Vigo Place, Aldridge, Walsall, West Midlands WS9 8UG)

Perkin Elmer Corp, 761 Main Ave., Norwalk, CT 06859-0012, U.S.A. (Perkin Elmer Ltd, Maxwell Road, Beaconsfield, Buckinghamshire HP9 1QA)

PL Thermal Sciences Ltd, Surrey Business Park, Kiln Lane, Epsom, Surrey KT17 1JF, U.K.

Rheometrics Inc., One Possumtown Road, Piscataway, NJ 08854, U.S.A. (Rheometrics U.K., Market House, 62 St. Judes Road, Englefield Green, Surrey TW20 0BU)

Seiko Instruments Inc., 6-31-1 Kameido, Koto-ku, Tokyo 136 Japan. (Stormage Scientific Systems Ltd., 2 Fleetsbridge Business Centre, Upton Road, Poole, Dorset BH17 7AA)

Setaram, 7 rue de l'Oratoire, 69300, Caluire, France. (Roth Scientific Co. Ltd., Roth House, 12 Armstrong Mall, The Summit Centre, Farnborough, Hants GU14 0NR)

Shimadzu Corp., 3, Kanda-Nishikicho 1-chrome, Chiyoda-ku, Tokyo 101, Japan. (V.A.Howe & Co. Ltd., Beaumont Close, Banbury, Oxon OX16 7RG)

Solomat Instrumentation, Glenbrook Industrial Park, 652 Glenbrook Road, Stamford CT 06906, U.S.A. (Solomat Mfg Ltd., 2 St. Augustines Parade, Bristol BS1 4XJ)

TA Instruments Inc., 109 Lukens Drive, New Castle, DE 19720, U.S.A. (Marlin Scientific Ltd, PO Box 98, Coventry CV1 4LP)

Subject Index